About the Author

Bill Phillips is president of the International Association of Home Safety and Security Professionals. He currently works as a security consultant and freelance writer whose articles have appeared in *Consumers Digest*, *Home Mechanix*, *Keynotes*, *Safe & Vault Technology*, *Security Dealer*, the *Los Angeles Times*, and many other periodicals. Mr. Phillips is also the author of 13 security-related books, including *The Complete Book of Locks and Locksmithing, Sixth Edition*;  *The Complete Book of Home, Site and Office Security*; and *The Complete Book of Electronic Security*, available from McGraw-Hill.

Master Locksmithing

**An Expert's Guide to Masterkeying, Intruder Alarms,
Access Control Systems, High-Security Locks,
and Safe Manipulation and Drilling**

Bill Phillips

New York   Chicago   San Francisco   Lisbon   London   Madrid
Mexico City   Milan   New Delhi   San Juan   Seoul
Singapore   Sydney   Toronto

The McGraw·Hill Companies

Library of Congress Cataloging-in-Publication Data

Phillips, Bill, date.
  Master locksmithing : an expert's guide to masterkeying, intruder alarms, access control systems, high-security locks, and safe manipulation and drilling / Bill Phillips.
    p. cm.
  Includes index.
  ISBN 978-0-07-148751-1 (alk. paper)
  1. Locksmithing. 2. Security systems. I. Title.
TS520.P546 2007
683'.3—dc22

2007021499

McGraw-Hill books are available at special quantity discounts to use as premiums and sales promotions, or for use in corporate training programs. For more information, please write to the Director of Special Sales, Professional Publishing, McGraw-Hill, Two Penn Plaza, New York, NY 10121-2298. Or contact your local bookstore.

2 3 4 5 6 7 8 9 0   DOC/DOC   1 2 1 0

ISBN 978-0-07-148751-1
MHID 0-07-148751-4

This book is printed on acid-free paper.

Sponsoring Editor Cary Sullivan	**Proofreader** Paul Tyler
Acquisitions Coordinator Alexis Richard	**Indexer** Karin Arrigoni
Editorial Supervisor David E. Fogarty	**Production Supervisor** Pamela A. Pelton
Project Manager Patricia Wallenburg	**Composition** TypeWriting
Copy Editor Marcia Baker	**Art Director, Cover** Jeff Weeks

Information contained in this work has been obtained by The McGraw-Hill Companies, Inc. ("McGraw-Hill") from sources believed to be reliable. However, neither McGraw-Hill nor its authors guarantee the accuracy or completeness of any information published herein, and neither McGraw-Hill nor its authors shall be responsible for any errors, omissions, or damages arising out of use of this information. This work is published with the understanding that McGraw-Hill and its authors are supplying information but are not attempting to render engineering or other professional services. If such services are required, the assistance of an appropriate professional should be sought.

To my son Michael

Contents

Foreword

The knowledge, skill, and expertise of any true professional can often be gauged by the library of literature that he or she keeps and has read.

The locksmith profession is also often one of the most challenging and rewarding. Few things in life are nearer and dearer to oneself than the security of one's family and worldly possessions. Having the knowledge and ability to instill and incorporate security, eliminating the helpless sense of vulnerability, is what locksmithing is all about. However, to accomplish that task, knowledge of all the available security devices and an intimate knowledge of their operation and applications are necessary.

Bill Phillips is a knowledgeable locksmith and talented writer. He is one of those individuals who encompasses the rare ability to make a wide variety of complicated topics enjoyable to read and easy to understand. That's a sure sign of a talented, versatile writer.

As the editor of the *National Locksmith* magazine, Bill Phillips is a cherished contributor of security-related topics to this publication. The *National Locksmith* is the oldest and most respected locksmith trade journal in the industry, serving locksmiths and security professionals since 1929. This monthly magazine has a group of the most experienced and well-respected writers in the field of locksmithing. In 2004, Bill Phillips became one of the distinguished elite to join the *National Locksmith*'s outstanding team of contributing writers, and I couldn't be more excited to have him.

Bill Phillips and the *National Locksmith* share one common goal—to provide informative, timely, useful, and cutting-edge information to beginning and experienced locksmiths. Bill Phillips accomplishes that through his writings and personal presentations. The *National Locksmith* does it through its books, software, the *National Locksmith* magazine, the *Institutional Locksmith* magazine, its web site at www.TheNationalLocksmith.com, and membership organizations, such as The National Safeman's Organization (NSO) and The National Locksmith Automobile Organization (NLAA).

Few people have as much knowledge or experience writing about locks or locksmithing as Bill Phillips does. As an author he has written for most of the locksmith trade journals, many general consumer magazines, the *World Book Encyclopedia* (the "Lock" article), and he has been published by one of the most respected and distinguished publishing houses—McGraw-Hill. Those are claims and accomplishments few others can make.

Greg Mango

Greg Mango is a 21-year veteran of the locksmith industry. He is currently the editor of the *National Locksmith* magazine and director of The National Locksmith Automobile Organization (NLAA). Greg has authored numerous security-related articles for multiple publications over the last 15 years, and his monthly "Mango's Message" editorial column is read by thousands.

Introduction

For about the last 20 years, I've been writing books and articles about locksmithing, including *The Complete Book of Locks and Locksmithing, Sixth Edition* and *Locksmithing and Professional Locksmithing Techniques, Second Edition*, both published by McGraw-Hill.

Whether you're a locksmithing student or apprentice, or just want to be a better locksmith, this book can help you. It can also help you prepare for various certification and registration tests, including the certified protection professional (CPP), registered professional locksmith (RPL), and registered security professional (RSP) tests. A copy of the RPL test is in Appendix F.

Another benefit of reading this book is that you can receive $10 off of your annual membership to ClearStar.com—one of the largest and most helpful websites for locksmiths. Just tell Jay (the webmaster) that you learned of the site by reading *Master Locksmithing* by Bill Phillips.

If you have questions or comments about this book, or have a book idea, I'd love to hear from you. You can write to me at Box 2044, Erie, PA 16512-2044. Or you can email me at LocksmithWriter@aol.com.

Basic Types of Locks and Keys: A Refresher

A locksmith needs to be familiar with all basic types of locks and keys. Because this book is mainly for apprentices and practicing locksmiths, I won't spend a lot of time on the basics. Review this chapter if you need a refresher.

A *lock* is a device that incorporates a bolt, shackle, or switch to secure an object—such as a door, drawer, or machine—to a closed, opened, locked, or an off or on position, and that provides a restricted means of releasing the object from that position. If anyone can just turn the door knob and walk in, there's no restriction. That's why a set of door knobs isn't a lock, but a key-in knob is.

Terms such as "mortise bit-key lock" and "Medeco key-in-knob lock" mean little to most people, but they provide useful information to locksmiths. Like other trades, locksmithing has its own vocabulary to meet its special needs.

Terminology

Laypersons often use a generic name like padlock, automobile lock (see Figure 1.1), or cabinet lock. Such names have limited value to a locksmith because they are so general. Those types of names simply refer to a broad category of locks that are used for a similar purpose, share a similar feature, or look similar to one another.

Figure 1.1   Common automobile locks.

Locksmiths identify a lock in ways that convey information needed to purchase, install, or service it. The name they use is based not only on the purpose and appearance of the lock, but also on the lock's manufacturer, key type, installation method, type of internal construction, function, and finish. To see a list of finishes, see Appendix A.

Some of the commonly used generic lock names include automobile locks, bike locks, ski locks, cabinet locks, gun locks, deadbolt locks, and patio door locks. See Figure 1.2. Sometimes generic terms have overlapping meanings. A padlock, for instance, can also be a combination lock.

The names used by a locksmith are typically formed by combining several words. Each word used provides important information about the lock. The number of words a locksmith uses for a name depends on how much information they need to convey.

When ordering a lock, for instance, the locksmith needs to use a term that identifies the lock's purpose, manufacturer,

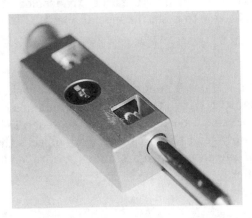

Figure 1.2   A patio door lock.

key type, appearance, etc. However, a name that simply identifies the lock's internal construction may be adequate for describing a servicing technique to another locksmith.

A *rim lock* is a lock that's designed to be mounted on the surface of a door or object. The interlocking deadbolt (or "jimmy proof deadlock") shown in Figure 1.3 is one type of rim lock. A *mortise lock* is installed in a hollowed-out (or mortised) cavity. A *bored* (or *bored-in*) *deadbolt* is installed by cross-boring two holes—one for the cylinder and one for the bolt. See Figures 1.4 and 1.5. A lock that uses a pin tumbler cylinder, for example, is usually called a *pin tumbler lock* or a *pin tumbler cylinder lock*. Figures 1.6 and 1.7 show some typical pin tumbler cylinders.

A lock that relies mainly on wards inside its case for its security is called a *warded lock*, as Figure 1.8 shows. Warded, lever tumbler, and pin tumbler are names that describe a lock's internal construction.

The *key-in-knob lock* refers to a style of lock that is operated by inserting a key into its knob, as seen in Figure 1.9. As a general rule, key-in-knob locks aren't very secure. They can be opened by knocking off the knob and using a screwdriver. That's why when installed on an exterior door, a deadbolt is

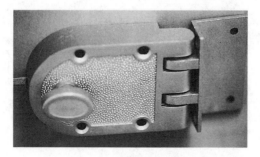

Figure 1.3   A jimmy-proof deadlock and strike.

Figure 1.4   A bored-in deadbolt lock (courtesy of Marks USA).

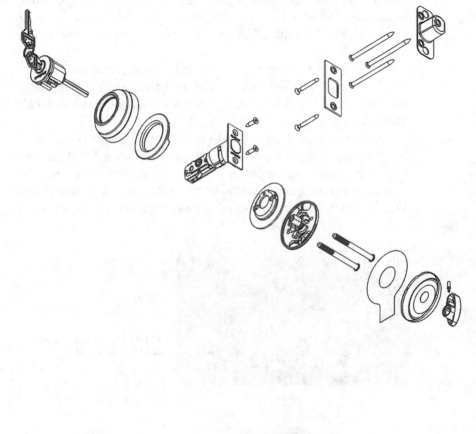

Figure 1.5 An exploded view of a bored-in deadbolt (courtesy of Kwikset Corporation).

4

Figure 1.6 Two pin tumbler cylinders.

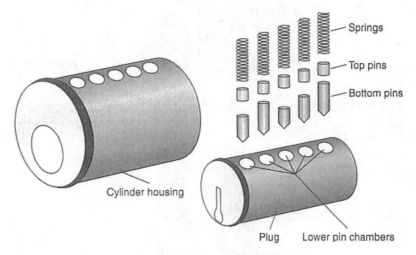

Figure 1.7 An exploded view of a pin tumbler cylinder.

Figure 1.8 A rim bit-key lock.

usually installed with the key-in-knob. That allows the security of a deadbolt with the convenience of a key-in-knob. See Figure 1.10.

A *lever lock* has a lever as a handle (see Figure 1.11). A *handleset lock* has a built-in grip handle. A *deadbolt lock* projects a deadbolt. As their names imply, the automobile lock, bike lock, ski lock, patio door lock, etc., are based on the purposes for which the locks are used. Sometimes, locks that share a common purpose look very different from one another.

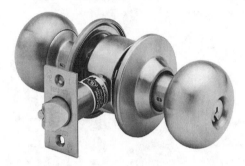

Figure 1.9   A key-in-knob lock (courtesy of Marks USA).

Figure 1.10   Typically key-in-knob locks for exterior doors are installed with a deadbolt.

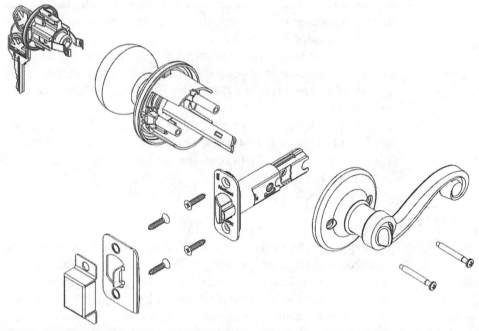

Figure 1.11 An exploded view of a lever lock (courtesy of Kwikset Corporation).

Manufacturers' Names

Locksmiths often refer to a lock by the name of the manufacturer, especially when all or most of the company's locks share a common characteristic. Locks manufactured by Medeco, for instance, all have similar internal constructions. Simply by knowing the lock is a Medeco Biaxial, a locksmith can consider options for servicing it.

Several lock manufacturers are so popular in the locksmithing industry that every locksmith is expected to be familiar with their names and the common characteristics of each manufacturer's locks. Those companies include Arrow, Best, Kwikset, Medeco, Schlage, and Weiser.

Key Types

Many times, a lock is identified by the type of key used to operate it. Bit-key locks and tubular key locks are two common examples. *Tubular key locks*, sometimes called *Ace locks*, are used on vending machines and coin-operated washing machines. *Bit-key locks* are used on many old closet and bathroom doors. When speaking about a bit-key lock, locksmiths usually use a term that reveals how it is installed—such as a surface-mounted bit-key lock.

Lock names based solely on key type don't indicate the lock's purpose, installation method, function, or appearance. They only refer to the type of cylinder or parts inside the lock case or housing.

Some people use the terms "lever lock" and "lever tumbler lock" synonymously. However, the latter refers to a type of internal construction, whereas the former refers to the type of handle used.

Warded Locks

A *ward* is a fixed projection designed to obstruct unauthorized keys from entering or operating a lock. One type of warded lock comes in a metal case, has a large keyhole, and is operated by a bit-key (commonly called a "skeleton key"). Such locks, called *bit-key locks*, come in mortised and surface-mounted styles, and are often used on closet and bathroom doors. Some inexpensive padlocks are also warded. They can be identified by their wide sawtoothlike keyways, and their squared cut keys. Warded locks provide little security because wards are easy to bypass.

Tumbler Locks

Tumblers are small objects, usually made of metal, that move within a lock cylinder in ways that obstruct a lock's operation until an authorized key or combination moves the tumblers into alignment. Several types of tumblers exist. They come in a variety of shapes and sizes, and they move in different ways. Because tumblers provide more security than wards, most locks today use tumblers instead of, or in addition to, wards.

A typical key-operated tumbler lock consists of a cylinder case, a *plug* (the part with a keyway), springs, and tumblers. The springs are positioned in a way that makes them apply pressure to the tumblers. The tumblers are positioned so when no key is inserted, or when the wrong key is inserted, the spring pressure forces one or more of the tumblers into a position that blocks the plug from turning.

Figure 1.12   A thumb turn is often used on the interior side of a single-cylinder lock (courtesy of Kwikset Corporation).

A lock can have more than one cylinder. A key-operated, single-cylinder lock has a cylinder on one side of the door (usually the exterior side), so that no key is needed to operate it from the other side. Typically, this lock can be operated from the noncylinder side by pushing a button or turning a knob, a handle, or a turn piece. See Figure 1.12. Key-operated, double-cylinder locks require a key on both sides of the door. Many local building and fire codes restrict the use of double-cylinder locks on doors leading to the outside because these locks can make it hard for people to exit quickly during a fire or other emergency.

Types of Tumbler Locks

There are three basic types of tumblers: lever, disc, and pin. Most *lever tumbler locks*, such as those used on luggage and briefcases, offer a low level of security. But, the lever tumbler locks used on bank safe-deposit boxes offer a high level of security. *Disc tumbler locks*, which offer a low-to-medium level of security, are often used on desks, file cabinets, and automobile doors and glove compartments. *Pin tumbler locks* can provide low, medium, and high security but, in general, they provide more security than other types of tumbler locks. Many prison locks and most house locks designed for exterior doors use pin tumbler cylinders.

A type of pin tumbler lock, called a tubular key lock (or tubular lock), has its tumblers arranged in a circular keyway. A *tubular key lock* uses a tubular key to push the tumblers into alignment. Because of its odd appearance, a tubular key lock is harder for most people to pick open, but locksmith supply houses sell picks for such locks. Tubular key locks are often found on vending machines, laundromat equipment, and bicycle locks.

High-Security Pin Tumbler Locks

A *high-security lock* is one that provides a high level of security to most types of attack, including drilling, sawing, wrenching, picking, and bumping. In most cases, such a lock will be Underwriters Laboratories (UL) listed and much more expensive than most locks. It will have two or more means of protection—such as hardened drill-resistant inserts, wards, sidebars, and patented keys. Medeco, ASSA, and Mul-T-Lock are a few brands of high-security locks. Figures 1.13 and 1.14 show two popular high-security keys.

Many lock manufacturers offer locks in several grades referred to as "light duty," "residential," "heavy duty," and "commercial." The grade descriptions are useful guidelines, but there are no industry standards for manufacturing each grade. One manufacturer's heavy duty, for example, may be of lower quality than another manufacturer's light duty. The grade names are meaningful only for comparing locks made by the same manufacturer.

A better measure of a lock's quality is the rating given to it by the American National Standards Institute (ANSI). The

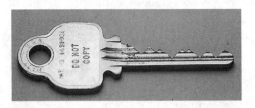

Figure 1.13   A Medeco angularly bitted key.

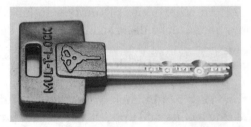

Figure 1.14   A Mul-T-Lock dimple key.

Good Lock Checklist

1. Bolt with at least one-inch throw.
2. Bolt made of hardened steel or with hardened steel insert.
3. UL listed.
4. Builders Hardware Manufacturers Association (BHMA) Grade 1 or 2.
5. Uses a patented key.
6. Requires two or more actions to open (such as pins reaching the shear line at the same time a sidebar moves into place).

three common ANSI standards for locks—Grade 1, Grade 2, and Grade 3—are based on a range of performance features. ANSI rates how well the locks resist forceful tests and whether their finishes hold up well, over extended periods of time. When you see one of these ANSI ratings, this is what you should know:

- *Grade 1 locks* are for heavy-duty commercial uses and would be overkill for most homes.
- *Grade 2 locks*, although designed for light commercial uses, can provide good protection for most homes.
- *Grade 3 locks* are for light residential uses.

Most locks sold to homeowners are either Grade 3 or have no ANSI rating. Only a few manufacturers—such as Kwikset Corporation and Master Lock Company—offer lines of Grade 2 locks through department stores and home improvement centers. These ANSI-graded locks often cost little more than locks that have no ANSI classification.

Should you sell a lock that doesn't have an ANSI classification? Aside from the directions on how to install it, you can't rely on information the manufacturer has printed on the packaging. Much of the description of the lock is just advertising hype. To be able to separate the hype from meaningful information, you need to rely on ANSI ratings and UL listings. When someone comes to you to buy a lock, you need to educate them about such standards.

Chapter

2

Picking Pin
Tumbler Locks

This chapter explains how to open standard pin tumbler locks, but high-security locks may require different tools and techniques.

With practice, you should be able to pick most standard pin tumbler locks open within a few minutes. In theory, any mechanical lock operated with a key can be picked because tools and techniques can be fashioned to simulate the action of a key.

A simple way to describe lock picking is to insert a lock pick and a torque wrench into a lock plug in such a way that the pick lifts each tumbler into place (where the right key would place them) while the torque wrench provides the pressure to turn the plug into the open position. You have to vary the pressure of the torque wrench. If you turn it too hard, the pins won't move into position. If you turn it too lightly, the pins won't stay in place. Usually, the problem is the torque wrench is turned too hard. Figure 2.1 shows someone picking a lock.

Lock picks come in a wide variety of shapes and sizes. Some are for specific locks, such as for tubular key locks. See Figure 2.2.

The key to picking locks fast is to focus on what you're doing and to visualize what's happening in the lock while you're picking it. If you know how pin tumbler locks work, it's easy to understand the theory behind them.

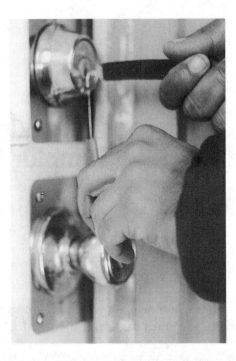

Figure 2.1   Most pin tumbler locks can be picked open with a pick and a torque wrench.

Figure 2.2   Lock picks come in a variety of shapes and sizes.

The slight spaces and misalignments between pins within a cylinder are what allow a lock to be picked. When locks are manufactured, there must be room in each of the lower and upper pin chambers (the holes along the length of the plug and cylinder housing) to allow the pins to easily move up and down in each pin chamber. If the pins were squeezed in too tight to move, the lock wouldn't work. Likewise, there has to be space between the plug and cylinder housing, so the plug can be turned to the locked and unlocked position. Lock picking takes advantage of those necessary spaces and slight misalignments in a lock.

In the locked position, the upper pin chambers are roughly aligned with the lower pin chambers, and one or more pins are lodged between the upper and lower pin chambers—obstructing the plug from turning. The correct key has cuts designed to move all the pins to the shear line at once, which frees the plug to be turned in a standard pin tumbler lock.

The upper and lower pin chambers aren't drilled in a perfectly straight line across the plug and cylinder. The drilled holes form more of a zigzag pattern. See Figure 2.3. In all but the most expensive locks there is a lot of play for the pins within the cylinder. That's because manufacturing processes aren't perfect.

When you use a pick to lift a bottom pin, you're also lifting the corresponding top pin (or top pins, in the case of a master-keyed lock). As a pin stack is being lifted, and you apply slight torque pressure to the plug with a torque wrench, the top of the bottom pin and the bottom of the top pin both reach the shear line at some point. See Figures 2.4 and 2.5.

The top pin goes into its upper pin chamber and leans against its chamber wall. The bottom pin, depending on its length, either stops at the shear line or falls back into the plug.

When a pin stack reaches the shear line, it turns a little, and a small ledge is created on the plug. That ledge, along with plug torque pressure, prevents the top pin from falling back into the lower pin chamber when the pick is removed. The top pin will stay on the plug's ledge at the shear line as long as adequate pressure is applied to the torque wrench.

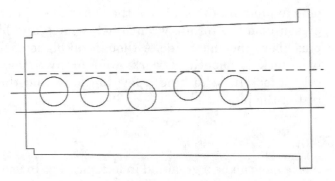

Figure 2.3   The holes along a lock plug aren't perfectly aligned.

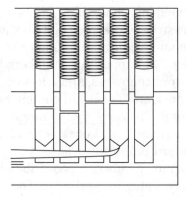

Figure 2.4   A lock pick is used to lift pin stacks, while a torque wrench provides turning pressure.

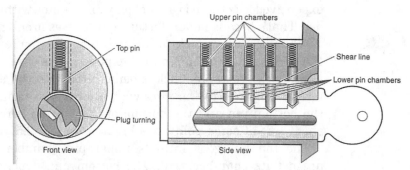

Figure 2.5   When the top pins and bottom pins reach a lock's shear line, the lock is in position to be turned.

A common method of picking locks is the rake method, or raking. To *rake* a lock, insert a pick (usually a half diamond or rake) into the keyway past the last set of pin tumblers, and then quickly move the pick in and out of the keyway in a figure-eight movement, while varying tension on the torque wrench. The scrubbing action of the pick causes the pins to jump up to (or above) the shear line, and the varying pressure on the torque wrench helps catch and bind the top pins above the shear line. Although raking is based primarily on luck, it sometimes works well. Many times, locksmiths rake a lock first, to bind a few top pins, and then pick the rest of the pins.

Lock Pick Gun

A *pick gun* can be a great aid in lock picking. To use a pick gun, insert its blade into the keyway below the last bottom pin.

Tips for Buying Picks

1. Buy a good selection of picks of various configurations.
2. Buy the thinnest picks you can find.
3. Get picks with comfortable handles.

Hold the pick gun straight, and then insert a torque wrench into the keyway. When you squeeze the trigger of the pick gun, the blade slaps the bottom pins, which knock the top pins into the upper pin chambers. Immediately after each squeeze, vary the pressure on the torque wrench. You will likely capture one or more upper pins in their upper pin chambers and set them on the plug's ledge. Then you can pick each of the remaining pin sets, one by one.

Before attempting to pick a lock, make sure the lock is in good condition. Turn a half diamond pick on its back, and then try to raise all the pin stacks together. Then, slowly pull the pick out to see if all the pins drop, or if one or more of the pins are frozen. If the pins don't all drop, you may need to lubricate the cylinder or remove foreign matter from it.

Hold the pick as you would hold a pencil—with the pick's tip pointing toward the pins. With the other hand, place the small bent end of a torque wrench into the top or bottom of the keyway, whichever position gives you the most room to maneuver the pick properly. Make sure the torque wrench doesn't touch any of the pins. Use your thumb or index finger of the hand that's holding the torque wrench to apply light pressure on the end of the torque wrench in the direction you want the plug to turn.

While using a pick, carefully lift the last set of pins to the shear line, while applying slight pressure with the torque wrench. The *shear line* is the space between the upper and lower pin chambers. Take a mental note of how much resistance you encountered while lifting the pin stack. Release the torque wrench pressure, letting the pin stack drop back into place. Then, move on to the next pin stack and do the same thing, keeping in mind which pin stack offered the most resistance. Repeat that with each pin stack.

Next, go to the pin stack that offered the most resistance. Lift the top of its bottom pin to the shear line, while varying pressure on the torque wrench. Apply enough pressure on the torque wrench to hold that picked top pin in place. Then gently move on to the next most-resistant stack. Continue lifting each pin stack (from most resistant to least resistant) to the shear line. As you lift each pin stack into place, you are creating a larger ledge for other top pins to rest on. When all the top pins are resting on the plug, the plug will be free to turn to the unlocked position.

No amount of reading will make you good at picking locks. You need to practice often, so you develop the sense of feel. You need to learn how to feel the difference between a pin tumbler that has been picked (that is, placed on the ledge of the plug) and one that is bound between its upper and lower chambers.

To practice lock picking, start with a cylinder that has only two pin stacks in it. When you feel comfortable picking that, add another pin stack. Continue adding pin stacks until you can at least pick a five-pin tumbler.

When you're practicing, don't rush. Take your time, and focus on what you're doing. Always visualize the inside of the lock and try to picture what's happening while you're picking the lock. For the best results, practice under realistic circumstances. Instead of sitting in a comfortable living room chair trying to pick a cylinder, practice on locks on a door or on a display mount.

With a lot of focused practice, you'll find yourself picking all kinds of locks faster than ever.

3

Bumping Keys

One of the most controversial subjects among locksmiths is whether or not to alert the public about a lock-opening technique called "key bumping." This technique is controversial because, unlike lock picking and key impressioning, key bumping is an easy technique to learn and it's effective on most pin tumbler locks. First, I present the main arguments for and against publicly disclosing information about lock bumping. Then, I explain how it's done. I decided to include the information in this book because I think all locksmiths need to understand it.

This chapter explains how to open standard single-action pin tumbler locks, but high-security cylinders that require two or more actions to operate may need different techniques and tools.

On August 28, 2006, the Associated Locksmiths of America (ALOA) issued the following press release:

"The Associated Locksmiths of America says that consumers have been unduly alarmed by attention-seeking individuals who have used the media to create a panic over recent reports of the use of 'Bump Keys' to commit burglaries. However, because of the widespread distribution of this information through various media, it now has the potential to become a real security threat to consumers.

The technique of using 'Bump Keys' is one of many methods used by locksmiths over the past 75 years to open locks for which there is either a cylinder malfunction or lost key. There are, in fact, many ways to prevent this method of opening so it is not a first-line technique that is used by professional locksmiths.

Initially, the individuals who promoted this information to the press may have been making what ALOA perceives as a misguided attempt at consumer 'awareness.' They gave the impression that opening locks by 'Key Bumping' was a widespread problem. It certainly has not been a method used by most burglars for many reasons. However, now that this method of opening some locks has become a popular theme, the most probable effect will be to stimulate the interest of would-be burglars to 'Bump Open' locks!"

Response to the ALOA Press Release

By Marc Weber Tobias

On August 28, ALOA posted a press release regarding recent publicity about the vulnerability of pin tumbler locks through the use of the bump key. Although not named, ALOA was clearly pointing the finger at myself and my associates who have made public the security issues from bumping that affect most mechanical cylinders, including those employed by the U.S. Postal Service and Mail Boxes Etc. Although I have always supported the goals of the organization, because of the position taken by ALOA, I felt obligated to respond.

From their brief statement, ALOA evidently believes the following to be true:

I and others (including some locksmiths) have made information public that was heretofore secret and unknown to the general public;

We made statements that burglaries have resulted from the use of bump keys in an effort to scare and "unduly alarm" the public;

The public does not need to know about bumping, nor were they at risk prior to the public disclosure;

Bumping now poses a serious threat to security but prior to the media coverage, it did not;

The recent publicity will only serve to educate criminals and does not serve any other legitimate purpose;

No locksmith or member of ALOA should be making any public statements about bumping and why it is a security threat.

ALOA clearly believes that "security through ignorance" should be the rule. If nobody knows about a vulnerability, then it does not pose a threat. Evidently, if we "kill the messenger" that will surely take care of the problem! Unfortunately, the criminals have known about bumping for quite some time, as have the sports lock picking groups. The vast majority of consumers were not aware of the insecurity of their locks. Amazingly, some manufacturers were also unaware of the vulnerability. I just met with one of the largest lock makers. They stated that they had no knowledge of bumping until they saw the news reports and read articles on the Internet. Some manufacturers have publicly attacked the concept of bumping, stating that it does not work on their locks, notwithstanding multiple reports and videos to the contrary. "Smoke and mirrors" is how one leading high security lock manufacturer described bumping. To make such blanket statements, by any manufacturer, is arrogant, denotes a lack of knowledge of the subject, is deceptive and misleads the consumer.

Perhaps the leadership of ALOA and some lock manufacturers might want to come up to speed on the new method of bumping. When reporters and kids can open the cylinders that locksmiths sell, there is a problem. Everyone that relies upon locks has a right to understand it so they can assess their own risk and take the appropriate steps. If we follow the ALOA logic, they and the locksmiths and security professionals are the only ones that should understand the problem and the inherent risks. Unfortunately, the vast majority of lock users do not get to deal with these experts, but have to make their own decisions.

The Facts

Since 2004, there has been a significant amount of publicity in Europe and on the Internet about bumping, including many videos from all over the world showing how to open locks. I was interviewed on our statewide CBS affiliate in 2004 and 2005 with regard to bumping and the vulnerability of post office and UPS mail box locks. In addition, I wrote a detailed article on the subject in Keynotes in 2005, as well as covering the subject in my book.

In December, 2005, I began consulting with the Postal Inspection Service regarding the vulnerability of their locks to bypass by bumping. They were not aware of the seriousness of the problem prior to my initial meeting with them and immediately escalated the matter to the highest levels in Washington. I believed this to be a serious problem that needed to be urgently addressed due to the increased publicity that bumping was receiving. I waited four months before publishing a report in order to give the postal service time to respond.

They did not request that I not publish my report. In fact, some management-level employees advocated that I should make the findings public so that enough attention would be drawn to the issue that something could be accomplished in Washington. I also recommended and continue to advocate that the postal laws be changed to prevent the trafficking in pre-cut bump keys. The Postal Service has issued a statement indicating that they have identified several security vulnerabilities and are addressing them, and have also begun replacing all post office box locks with a new design. That is a direct result of the media attention to the subject and clearly serves the public interest.

In March, 2006, a detailed report was issued by Consumer Reports in the Netherlands. Their findings examined the test results that were obtained in evaluating about seventy lock manufacturers' products. This was a joint effort between the police, Consumer Reports and the Dutch sports lock picking group, known as TOOOL. That article stated that a majority of the locks could be opened without difficulty, even some with high security ratings. As a result of that publication and after consulting with a number of manufacturers in Cologne, I posted a White Paper on www.security.org that detailed the real threat from bumping and the legal issues involved.

In July and August, I lectured in New York and Las Vegas at the international hacker conventions. According to ALOA, these are gatherings of criminals and persons of "questionable character." The fact is, most attendees are corporate IT professionals, security managers and government agents. At Defcon in Las Vegas, I lectured with Matthew Fiddler, a security expert employed by a Fortune 100 company. Barry Wels, one of the leading experts on bumping and the person who is most responsible for bringing this issue to light in Europe, co-presented with me at the New York conference.

The security vulnerability of pin tumbler locks affects just about everyone and it did not take the news media long to figure it out, especially when a young girl demonstrated opening a popular five pin cylinder in seconds with no prior experience. Now, many locksmiths are speaking out and acknowledging the problem and working to fix it. In my view, this is the responsible thing to do.

I challenge ALOA to produce one article or press release that stated that criminals had utilized bumping to effect entry! The media has asked for such information, but it has not happened in any widespread fashion, even with the publicity during the past two years. Bumping is a real threat, but there is a remedy: just install better locks. Nobody has said that there is an outbreak of burglaries, but there surely could be, and that is precisely the issue. Why is ALOA so concerned about letting the public in on their "secret"? Maybe it is because just about everyone is affected and they can understand the simplicity of the attack and thus its potential danger.

Does ALOA really believe that we should have waited for the criminals to deploy bumping as a popular and common method of entry before we warned the public of the threat? Who would that policy place at risk? Does ALOA actually advocate trying to keep security vulnerabilities secret when it affects millions of people, hoping that nobody will find out? This is not the kind of problem that the manufacturer can easily remedy, especially in the hundreds of millions of locks that are already installed. So should we place everyone at risk, or should we give them the opportunity to opt for more security and upgrade their locks?

Just how would ALOA go about warning the consumer to even give them that option? If this was a vulnerability that was in a product that did not affect a large segment of the population, then I would say to let the manufacturers quietly do a recall or fix the problem. But that course of action would not work in this case. So, there are only two options: keep quiet and allow widespread losses to occur that would place millions of people at risk, or warn them. And if we opted for the first alternative, keeping quiet, and the media learned of the vulnerability, then just how would ALOA, as the representative of the locksmith community, explain the fact that they knew about the potential security vulnerability for many years but failed to do anything about it? Their answer would surely be interesting!

The locksmith, in my view, should be proactive and suggest, where appropriate, an upgrade to better locks. Of course, there is a problem in doing this, as I am sure ALOA recognizes: the locksmiths would have to admit that they knew, but said nothing about the vulnerability in the locks that they have been selling. But then again, perhaps ALOA should be the one to respond to that issue, given their policy of non-disclosure of security defects to the end user.

The real question for ALOA is why they have not been pushing the lock manufacturers to deal with this problem, given that they have known about it for so long. The illogic is striking. If ALOA and their members have known that they have been selling and installing locks that could be easily bypassed, why would they continue to do so and place their customers at risk without warning them? The short answer, but not a good one, is that ALOA prevents its members from disclosing defects in any detail to the public. Why would that be? Surely it could not be linked to revenues received from those very same manufacturers and institutional organizations who are concerned with their embedded base of assets which could be at risk, to say nothing of the potential for lawsuits for negligence and product misrepresentation that could result!

For ALOA to state that the technique of bumping was not public information prior to July is untrue and they know it. Evidently, they believe that there are still secrets and that the public does not have a right or a need to know about vulnerabilities in the locks that they purchase and rely upon for security. The reality is that there are no more secrets! The Internet took care of all of that. This is not the eighteenth century with locksmith guilds, where information about locks was tightly controlled. This is the twenty-first century, where information about everything is instantly accessible. And if you really think that you could publish a general and vague warning about the security of pin tumbler locks but not specifics, it would take about twenty-four hours for detailed reports to start showing up on the Internet!

I believe that ALOA prevents locksmiths from disclosing specific security vulnerabilities to the public to their detriment. I have advocated, as a lawyer, that this is bad public policy, irrational, and will ultimately lead to liability on the part of both the locksmith and ALOA. The public relies upon the locksmith as their first line of security. If they sell cheap locks, like the eleven-year-old girl opened in seconds in Las Vegas, they have an obligation to warn the prospective purchaser of the risks in using such products. By doing that, they would be acting responsibly, meeting their legal obligations, and most importantly, fulfilling their ethical duty. I know some locksmiths disagree with me on this point, and they made their views known two years ago after I posted an editorial in Keynotes on the subject of liability and full disclosure. But at the end of the day, full disclosure is the best policy. An educated consumer makes for a better customer, and a more secure one.

ALOA's contention that the public does not have a right or a need to know is irresponsible and without logic. The public, not the locksmith, should be making security decisions based upon a full understanding of the risks, whether from bumping, the compromise of master key systems, or other simple methods of attack. This means that they should understand how easy or difficult it is to open a cylinder, then make their own judgment as to whether that cylinder provides sufficient security. A failure to disclose security vulnerabilities will subject the locksmith to civil liability for misrepresentation and negligence, should there be a loss or injury resulting from the failure of the locks that were recommended by him. I can

guarantee that ALOA would be joined in any such law-
suit, because the locksmith would look to them for com-
pensation, claiming that they were following the ALOA
mandated policies on disclosure.

Might I suggest that rather than attacking "misguided
individuals" for making a potentially serious problem
public, which ALOA has now admitted is a significant
security threat to everyone, they should be taking the
lead to deal with the real issues. Specifically, I would urge
them to:

Change their rules to allow locksmiths to educate the
public in security vulnerabilities of their products;

Form an industry-wide consortium of manufacturers to
improve the current technology to frustrate bumping;

Educate the consumer with regard to the availability of
high security locks;

Encourage and work with sport lock picking groups to
identify security vulnerabilities. It will help everyone,
and the fact is, these groups are now operating in
America. ALOA should view them as allies, not enemies.
The fact is, some of their members are professional locksmiths, safe technicians, and Fortune 100 Security
Professionals responsible for the protection of critical
financial assets in America;

Join me in proposing legislation to prohibit the sale of
pre-cut bump keys through interstate commerce.
Currently, postal regulations specifically exempt bump
keys from all such prohibitions. Many sites are now selling these keys and are placing everyone at risk;

Work with UL and other standards organizations to
insure that high security ratings encompass bumping;

Work with lock manufacturers and encourage them to
provide warnings on their product packaging that alerts
the public about security vulnerabilities in their locks.
The public needs to know what they are buying and the
attendant risks;

Work with major retailers such as Home Depot and
Lowe's to encourage them to only sell locks with appropriate warnings on their packaging. Consumers that purchase these locks do not generally have the benefit of
dealing with a locksmith, yet they need the information;

Propose legislation that makes the possession of bump keys by unauthorized individuals equivalent to the possession of burglary tools.

In my view, ALOA and every locksmith should recognize that bumping is perhaps the most efficient method of bypass of a conventional pin tumbler cylinder, and thus the most serious threat. Virtually every conventional pin tumbler lock is at risk. Why not address the issue head-on and educate the public to upgrade their locks, where warranted. The locksmith is the first line of defense, and bumping can provide a real opportunity to serve the public and enhance their security. They should embrace that potential, not attack those who have dared to bring this problem to light. Millions of pin tumbler locks were insecure long before I or my associates brought the issue to the public. Media attention has served the public interest. At least now, they understand their vulnerability and can choose to do something about it. If they elect to ignore the risk, that is their decision, but at least now they have the knowledge to make that judgment.

How to Bump Locks

Bumping locks requires a key that fits all the way into the plug and a tool to tap on the key bow. To prepare the key (or blank), you need to cut each space to its deepest depth. For many keys, that's the "9" depth—which is why some people call bump keys "999" keys. To make a bump key for a standard Kwikset lock, for example, you cut a "6" at each space because "6" is the deepest normal cut for the lock. See Figure 3.1.

You don't need a new key blank to make a bump key. You can use any key that goes all the way into the plug and uses standard-cut depths for the lock. See Figure 3.2. The easiest way to make a bump key is to use a code machine or a code key cutter. If you don't have one, ask a local locksmith to make bump keys for you. You can also make them with a file, a caliper, and depth and space information. To get the right depth and spaces for the cuts, you can use depth and space charts. For depth and space information for some popular locks, see Appendix B. The space charts show you where to place the cuts along the length of the key. The depth charts show you how deep to make the cuts.

Figure 3.1   A bump key has all of its cuts at the deepest cut for the lock to be opened.

Figure 3.2   To bump a key, you first need to fully insert the bump key into a lock's plug.

When making a bump key, be careful not to cut any space too deep. File a little metal off the shoulder stop of the key (about 0.25 inches) on the key's biting side. Don't cut too much off the key. You can always cut off a little more, but you can't add metal to a key.

The first step to using the bump key is to fully insert it into the lock's plug. Then, pull the key out until you hear or feel one click. Use one of your hands to provide turning pressure on the key bow, while simultaneously tapping the back of the bow with a screwdriver handle, a tool designed to bump keys, or a small hammer. Tap the key hard enough for it to go fully into the plug. See Figures 3.3 and 3.4.

If the plug doesn't turn, remove the key, and then reinsert it all the way. Pull the key out until you hear one click. Then, again apply turning pressure, while tapping the bow. Sometimes you may need to tap the bow several times to open a lock.

Most standard locks can be quickly bumped open, but the technique doesn't work on all locks—especially locks that require two or more actions to occur to open the lock. For example, locks that require pins to meet at the shear line,

Figure 3.3 You can bump a key by using your thumb to apply turning pressure to the key's bow while tapping the bow, pushing it fully into the cylinder.

Figure 3.4 You can use a bumping tool to bump a key into the lock.

while simultaneously requiring the movement of side pins, side bars, or rotating elements.

Bumping keys works similarly to picking a lock with an electric or manual pick gun. By hitting the key bow, the bottom pins slide up into the top of the cylinder to momentarily create a shear line, thus allowing the cylinder plug to turn if you have the right amount of turning pressure.

What to Do About Bump Keys

Bump keys are easily available over the Internet. Some sellers require customers to prove they're locksmiths before selling bump keys. But others will sell them to anyone.

To help prevent the problem of burglars using bump keys, make sure you don't make them for anyone other than another locksmith. And don't sell key blanks to anyone other than locksmiths.

The widespread concern about the use of bump keys doesn't have to be a negative thing for locksmiths. If a customer is concerned about it, you can try to sell them a more secure lock. Many locksmiths categorize their locks as good, better, and best security, and then they let the customers choose.

Tips for Bumping Keys

1. Use a code machine or depth and space charts to file the manufacturer's deepest cut at each space.

2. Shave a little off the shoulder of the bump key at the biting side.

3. Use one hand to provide turning pressure on the bow in the direction you want the lock to turn.

4. Fully insert the bump key.

5. Pull the bump key out one click.

6. Use a screwdriver handle or bump tool to tap the key a couple of times on the back of the bow, while maintaining turning pressure with your other hand.

7. If the lock doesn't open, remove the key and start again.

Impressioning
Locks

This chapter explains two things: what impressioning is and how to do it.

From the outside, *impressioning* is inserting a prepared key blank fully into a keyway (see Figure 4.1), and then twisting the blank clockwise and counterclockwise, in a way that leaves tumbler marks on the blank, which shows where to file the blank to make a working key. Figure 4.2 shows tumbler marks. You then file the marks, clean the blank, and reinsert it, and then twist it again and file at the new marks. At some point, you'll have a working key. With practice, you should be able to impression most pin tumbler locks within five to ten minutes.

Pin Tumbler Locks

To impression a pin tumbler lock, you first need to choose the right blank (one that fits fully into the keyway). If the blank is too tight, you won't be able to rock it enough to mark it. The blank also needs to be long enough to lift all the pins. If you use a five-pin blank on a six-pin cylinder, you probably won't be able to impression it because the sixth pin won't mark the blank. To choose the right size blank, use a probe or pick to count the number of pin sets in the lock.

The material of the blank needs to be soft enough to be marked by the pins, but not so soft as to break off while you're twisting or rocking the blank. Nickel-silver blanks are too hard for impres-

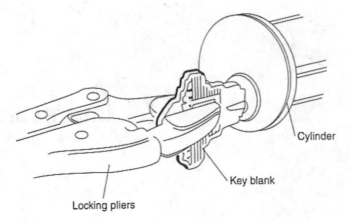

Figure 4.1   Insert the prepared blank fully into the cylinder.

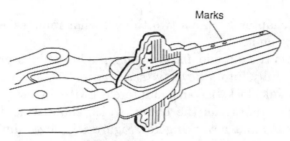

Figure 4.2   When you remove the blank, look for tiny smooth marks made by the lock's pin tumblers.

sioning because they don't mark well. Aluminum blanks are soft enough to mark well, but they break off too easily. Brass blanks work best. Nickel-plated brass blanks are also good for impressioning because the nickel plating can be filed off.

Filing the Blank

New key blanks have a hardened glazed surface that hinders impressioning, unless you prepare them. To prepare a blank, shave the length of the blade along the side that comes in contact with the tumblers. Shave the blank at a 45-degree angle without going too deep into the blank. See Figures 4.3, 4.4, and 4.5. You want the blank's biting edge to be sharp (a knife edge) without reducing the width of the blank. File forward only. Don't draw the file back and forth across the blank!

Figure 4.3   Use a file to pre-pare a key for impressioning.

Figure 4.4   File along the length of the blade.

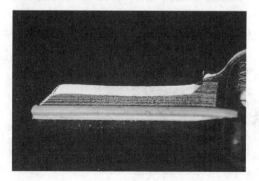

Figure 4.5   Shave the biting edge of the blank.

Use a round or pippin file with a Swiss No. 2, 3, or 4 cut. A coarser file will shave the blank quicker, but it will leave rougher striations on the blank—making it harder to see the pin marks. A finer file will make the marks easier to see, but it will clog quicker while you're filing. This makes impressioning take a long time. You probably won't find impressioning files at a hardware store or a home improvement center, but they're sold through locksmith supply houses. For a comprehensive list of locksmithing supply houses, see Appendix C.

Tips for Impressioning Locks

1. Use brass key blanks.
2. Shave the biting side of the blank to 45 degrees, without decreasing the width of the blank.
3. File in one direction, not back and forth.
4. Use good lighting to help you see the impression marks.

Another popular way to prepare the blank is to turn the blank over and shave the other side along the length where the tumblers touch. After shaving both sides of the biting edge at 45 degrees, you will have a double-knife edge.

Other Useful Equipment and Supplies

In addition to a file, you need a key-holding device, such as an impressioning tool or a 4- or 5-inch pair of locking pliers. See Figures 4.6 and 4.7. A magnifying glass can be helpful for seeing impressioning marks. A head-wearing type lets you see the marks and file the marks at the same time. See Figure 4.8. Although they aren't essential, you can use depth and space charts, as well as a caliper to file marks more precisely.

Figure 4.6 You can use pliers to hold the blank while impressioning.

Figure 4.7 An impressioning tool can be helpful for holding the key blank while impressioning.

Figure 4.8 A head-wearing magnifying glass leaves your hands free while impressioning.

5

High-Security Locks

There is no consensus among locksmiths about what makes a lock "high security." Some consider any lock that is UL listed as high security, but many of those can be easily defeated. Most locksmiths would agree that to be considered high security, a lock needs to provide a high level of resistance to all common methods used to defeat locks, such as wrenching, sawing, drilling, picking, and bumping. The most secure locks also provide a high level of key control. The harder it is for an unauthorized person to have a duplicate key made, the more security the lock provides. That's why high-security lock manufacturers use patented keys.

In most cases, a high-security lock requires two or more actions to occur simultaneously to operate the lock. For instance, the lock may require its pins to reach the shear line at the same time that a side bar moves into the right position.

Underwriters Laboratories Listing

Founded in 1894, Underwriters Laboratories (UL) is an independent, nonprofit product-testing organization. A UL listing (based on UL standard 437) is a good indication that a lock offers more than ordinary security. If a lock or cylinder has such a listing, you see the UL symbol on its packaging or on the lock. However, UL doesn't test locks for bump key resistance. Some of UL's requirements follow:

1. All working parts of the mechanism must be constructed of brass, bronze, stainless steel, or equivalent corrosion-resistant materials or have a protective finish complying with UL's Salt Spray Corrosion test.

2. The lock must have at least 1,000 key changes.

3. The lock must operate as intended during 10,000 complete cycles of operation at a rate not exceeding 50 cycles per minute.

4. The lock must not open or be compromised as a result of attack tests using hammers, chisels, screwdrivers, pliers, hand-held electric drills, saws, puller mechanisms, key impressioning tools, and picking tools.

The attack test includes ten minutes of picking, ten minutes of key impressioning, five minutes of sawing, five minutes of prying, five minutes of pulling, and five minutes of driving. Those are net working times, which don't include time used for inserting drill bits or otherwise preparing tools.

Key Patents

Not all patented keys offer a lot of key control. There are two relevant types of patents: design and utility. A *design patent* only protects how a thing looks; a *utility patent* protects how a thing works. In 1935, Walter Schlage received a design patent for a key bow. The bow's distinctive shape made it easy for locksmiths and key cutters to recognize the company's keys. However, the patent didn't prevent aftermarket key blank manufacturers from making basically the same key blanks with slightly different bow deigns. Utility patents offer more protection against unauthorized manufacturing of key blanks and locks.

All patents eventually expire, and they aren't renewable. Design patents are granted for 14 years. Utility patents last for 17 or 20 years. The original utility patent for Medeco Security Locks (No. 3,499,302) was issued in March 1970. Because the patent prevented companies from making key blanks that fit Medeco locks, the company was able to maintain maximum key control. As soon as the patent expired (in 1987), compatible key blanks were offered by many companies. Medeco responded by introducing a new lock, the Medeco biaxial, getting a utility patent (No. 4,635,455) for it in June 1987. That protected the key blank until June 2004.

Other important utility patents include ASSA (No. 4,393,673), granted July 1983; Schlage Primus (No. 4,756,177), granted July 1988; Kaba Gemini (No. 4,823,575), granted April 1989; and Kaba Peaks (No. 5,016,455), granted May 1991.

If key control is important to your customer, make sure you find out what type of patent a lock and key have, and when the patent was issued.

The rest of this chapter covers ASSA high-security locks, and is from the ASSA Technical Service Manual.

ASSA TWIN 6000

Technical Service Manual

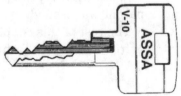

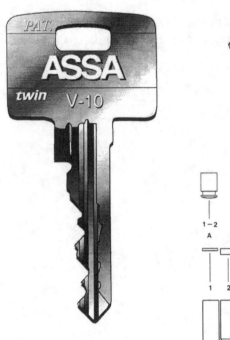

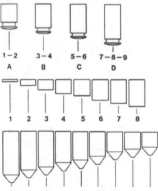

ASSA TWIN 6000

CUT AWAY DIAGRAM

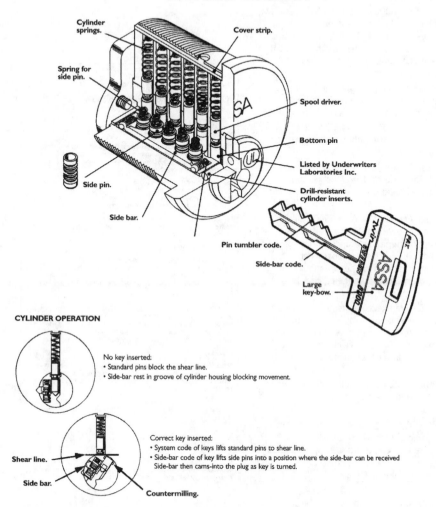

Cylinder springs.

Cover strip.

Spring for side pin.

Spool driver.

Bottom pin

Listed by Underwriters Laboratories Inc.

Side pin.

Drill-resistant cylinder inserts.

Side bar.

Pin tumbler code.

Side-bar code.

Large key-bow.

CYLINDER OPERATION

No key inserted:
• Standard pins block the shear line.
• Side-bar rest in groove of cylinder housing blocking movement.

Correct key inserted:
• System code of keys lifts standard pins to shear line.
• Side-bar code of key lifts side pins into a position where the side-bar can be received Side-bar then cams-into the plug as key is turned.

Shear line.

Side bar.

Countermilling.

ASSA TWIN 6000

THE UNPARALLELED SECURITY CYLINDER

The ASSA Twin 6000 cylinder is a high security option for any ASSA lock and retrofits most other manufacturers' locks. A specially shaped key simultaneously operates two independent locking mechanisms. Both the key and sidebar incorporate precision coding.

MATERIALS. Cylinder housing and inner plug are high-quality brass. Pins are stainless steel, hardened stainless steel or nickel silver.

MANUFACTURER TOLERANANCES. Precision machining insures smooth, positive operation and extended cylinder life.

DRILL RESISTANCE. In addition to side and tumbler pins, case-hardened drill-resistant inserts are embedded in the cylinder plug and housing.

PICK RESISTANCE. ASSA Twin 6000 cylinders have two independent shear lines that help assure optimum pick resistance. "False" grooves in the side pins catch the sidebar when improperly positioned. When rotational force is applied, countermilling in the cylinder plug catches the spool driver pins.

KEYS. ASSA Twin keys are manufactured from quality nickel silver material. The key's rounded back facilitates smoother operation and minimize wear. The large key bow simplifies key identification, and use by manually impaired people.

OTHER DESIGNS. ASSA Twin 6000 cylinders are available in different shapes for use in many types of locks.

PATENTS. Existing U.S patents and those existing and pending in over two dozen manufacture or use of unauthorized keys or cylinders. Any infringement of the ASSA patents rights will be forcefully prosecuted in court.

Products, equipment, finishes, models, specifications and availability are subject to change without notice.

HIGH SECURITY CYLINDERS FOR DEMANDING APPLICATIONS

Mortise and Rim Type Cylinders

ASSA 6000 mortise and rim cylinders are designed to replace original cylinders in many types of common locks. Made to standard dimensions and available in a variety of finishes, these reliable cylinders stand up to the heaviest use imaginable.

ASSA 6000 mortise cylinders will improve the security of locks installed in aluminum storefront doors, as well as standard mortise locks by Schlage, Yale, Sargent, and other manufacturers. They also can be applied to a wide variety of special locking devices.

Rim-type 6000 cylinders are designed for commercial exit and panic devices, as well as vertical deadbolt locks commonly used on apartment and condominium doors.

Key-In-Knob Cylinders

Knob locks by Schlage, Yale, Russwin, Sargent, and other manufacturers can be retrofitted with ASSA 6000 high security cylinders. Specifically engineered shapes replace only the original cylinder – greatly improving key control without the cost of replacing the entire lock!

High Security Deadbolts

ASSA 6000 deadbolt locks can be used to replace existing deadbolts or improve security on other doors. Two types of bolts and strike plates simplify installation in either wood or metal doors and frames, steel, saw-resistant deadbolt has full one inch throw. Two models are available: double-cylinder and single cylinder with inside thumbturn.

Padlocks

Three grades of ASS 6000 padlocks expand the application of master key systems: they also provide independent security on gates, machines, vehicles, etc. Case-hardened shackles and ball-locking mechanisms are housed in brass or steel barrels.

Rekeyable and virtually maintenance free, each ASSA 6000 padlock offers the same high standards of key control, pick resistance, and rill resistance common to all ASSA 6000 products.

*One or more of the following U.S. Patents covers ASSA 6000 cylinders, key blanks, and other components: 4,356.713 – 4.393673 – 4.577.479 – 274.302 – 278.880 – 264.680

ASSA TWIN 6000

DEPTHS

General Rules

A. No. 1 cut is deepest cut, No. 9 is shallowest.

B. Maximum adjacent cut of 5. Example: 1-6 cut acceptable, not 1-7.

C. Depths: Measure from bottom of key blade to bottom of the cut.

No 1 = 4.03 (.1587")	No 4 = 5.83 (.2295")	No 7 = 7.63 (.3004")
No 2 = 4.63 (.1823")	No 5 = 6.43 (.2531")	No 8 = 8.23 (.3240")
No 3 = 5.23 (.2059")	No 6 = 7.03 (.2768")	No 9 = 8.83 (.3476")
Cut depths tolerance	+0.00 -- -0.04 mm	
	+.000" -- -.0016".	

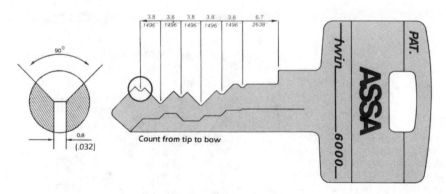

Count from tip to bow

General rules

A. 0.8 mm (.032") wide base.

B. 90° cutting angle.

C. System and KD code calculated from tip to bow.

An ASSA ABLOY Group Company

ASSA TWIN 6000

ASSA TWIN 6000 KEYS, KEYBLANKS

Key blanks

A. Sidebar Code is cut at ASSA Factory. Key blanks are provided as shown.

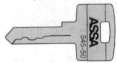

B. Sidebar code/profile can be identified on key bow.
Example 545-50

Key stamping

All new keys should be stamped as original. Stamping procedures are as follows.

	KD	MK
Gamma		2A295G 1AA
Beta		2B951B AA1
Delta		2C621D AC5
Alpha		2D586A B68

Decoder for KD/KA combinations.
Count from tip to bow.

Example: EAEGJC = 126958

	1	2	3	4	5	6
1	E	F	J	B	G	A
2	J	A	B	C	D	F
3	F	C	D	J	H	E
4	C	H	A	D	B	J
5	D	G	F	E	J	H
6	B	D	E	H	A	G
7	H	B	G	F	C	D
8	G	J	H	A	E	C
9	A	E	C	G	F	B

> KD Blind Codes can be found on factory code tags.

Key cutting

Sidebar code is pre-cut at the ASSA Factory. Pin Tumbler Codes can be cut on most code machines and duplicating machines. Manual duplicating machines, historically, produce better duplicating keys than automatic machines. The Twin 6000 key is a hefty quality key. ASSA recommends quality cutters. ASSA has cutter and code cards available for the HPC 1200 CN code machine.

ASSA TWIN 6000

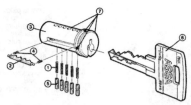

SIDE BAR COMPONENTS

1	Side pin springs
2	Side bar
3	Cylinder plug
4	Side bar springs
5	Side pins
6	Key
7	Drill resistant pins

Side bar assembly method is the same for all
ASSA Twin 6000 cylinders.

Side bar assembly method is for all
ASSA Twin 6000 cylinders

1. Turn cylinder plug upside down.
   Insert the side pin springs into the
   holes.
   NOTE: These pin holes are drilled
   from the bottom side of the cylinder
   plug.

2. Insert the side pins with the
   hollowed end towards the spring. Be
   sure that the springs and pins are
   properly seated in the recesses. If a

 spring or pin is jammed,
   the cylinder will not
   function.
   NOTE: All side pins are
   the same.

3. Keep the cylinder plug upside down
   and insert the key, for that cylinder,

 to keep the side
   pins in place.
   NOTE: Depress
   side pins in
   order to insert
   the key.

4. Insert the short springs (side bar
   springs) in the small holes at each

 end of the
   groove along
   the cylinder
   plug for the
   side bar. Be

sure that the springs seat correctly.

5. Place the side bar so that the shelves
   (lug) correspond to the deep waist
   of the side pins. Press the side bar
   into groove so that it can be fully
   depressed into the cylinder plug.
   NOTE: The key should be kept in the
   cylinder plug at all times during
   assembly.

 If side bar does not
   fully depress into
   the cylinder plug,
   side bar may be
   inverted.

6. Slide the cylinder plug into cylinder
   housing. If the position of the side
   bar is correct, the cylinder plug will

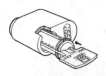 turn. Hold
   the cylinder
   plug in
   place when
   drawing
   out the key.

7. Fasten the tail piece and the tail
   washer. After assembly, test the

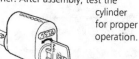 cylinder
   for proper
   operation.

An ASSA ABLOY Group Company

ASSA TWIN 6000

PIN TUMBLER ASSEMBLY

- When ordering complete or sub assembled cylinders please state the sidebar code and the keyway.
- Always turn key to right of clockwise when removing cylinder plug with following tool.
- It is recommended that some type of holding fixture be used to prevent the cylinder from tipping when plugging the pin chambers.

1. Closing plugs
2. Cylinder springs
3. Spool drivers
4. Master pins
5. Bottom pins
6. Cylinder housing
7. Cylinder plug
8. Tail piece
9. Tail piece washer
10. Screws
11. Drill resistant pin

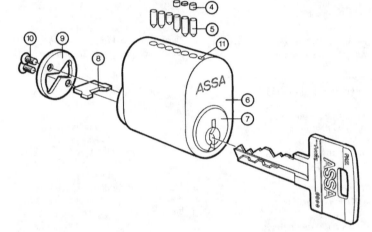

PIN TUMBLER CODE Example

Cylinder bitting

27838	= Bottom pins
314	= Master pins
ABDBAB	= Spool drivers

Key bitting

| 248424 | = Chance Key combination |
| 248738 | = Master key combination |

NOTE:
Always start from the tip of the key

Pin Tumbler system

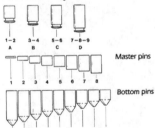

Master pins

Bottom pins

ASSA TWIN 6000

ASSA®

PIN TUMBLER ASSEMBLY

1. Insert the bottom pins as per the bitting sheet.

2. In case of Master-keyed cylinders, master pins are placed over the bottom pins as stated, according to bitting requirements.

3. Spool drivers must correspond to the length of the bottom pins and are then inserted onto the cylinder housing as follows:

Spool top pin A for bottom pin 1-2
"   B   "   3-4
"   C   "   5-6
"   D   "   7-8-9

Note: In pin chambers with bottom pin and master pins the total length of the bottom and master pins determines the length of the spool drives. (see example)

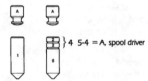

4. Cylinders springs are placed over each spool driver in the cylinder housing

5. Insert the closing plugs. Press the plugs down until they are flush with the top of the cylinder housing.

Note: Care should be taken not to damage the finish of the cylinder if hammering. Other methods of plugging apply to other types of ASSA cylinders.

6. After assembling, test the cylinder function for proper operation. For lubrication, use only ASSA spray.

ASSA TWIN 6000

1. Insert the side bar mechanism, bottom pins, master pins and springs with same method a described earlier. See the line drawing for the location of the components.

2. Fasten the tail piece and the tail piece washer with the tail screws.

3. After the cylinder assembly test for proper operation.

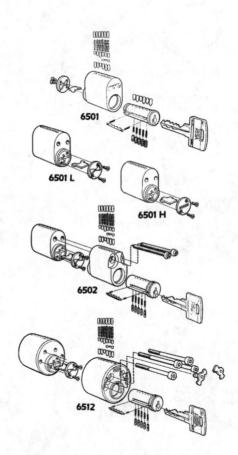

6501

6501 L

6501 H

6502

6512

ASSA TWIN 6000

MORTISE AND RIM CYLINDER ASSEMBLY

1. Insert the side bar mechanism, bottom pins, master pins, springs with method as described earlier.

2. Insert pin chamber closing strip into pin bible and then "stake" with a small chisel or similar tool so the strip will remain in position.

3. After the cylinder assembly test for proper operation.

a. The ASSA Mortise cylinder can be assembled utilizing one of seven different cams. See schedule.

b. The 1 1/8" (6551) cylinder housing is universal in that it is manufactured to work both as mortise or rim cylinder. The housing is both threaded mortise and tapped for rim screws. Mortise cams and rim tails have universal fittings.

c. The 1 ¼" (6552) cylinder housing is not tapped for rim installation.

d. Rim cylinder screws and mounting plate are available as an option.

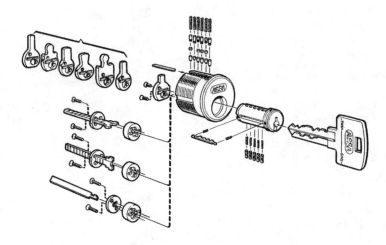

ASSA TWIN 6000

ASSA®

KEY-IN-KNOB ASSEMBLY

1. Insert the side bar mechanism, bottom pins, master pins and springs with the same method as described earlier.

2. Insert pin chamber closing strip into pin bible and then "stake" with a small chisel or similar tool so the strip will remain in position.

3. After cylinder assembly is completed, test for proper operation before installing in lockset.

65611 Schlage

1. Break off tailpiece provided to accommodate different functions. Cut to proper length at score.

2. The tailpiece can be located in either vertical or horizontal positions.

3. When installing the tailpiece – depress cylinder cap pin and tighten cap clockwise until is tight, then back off one or two notches.

Note: The cap must be properly adjusted. If the cap is too loose the key cannot be withdrawn because the cylinder plug and cylinder housing pin chamber will not line up properly. If the cylinder cap is too tight the cylinder will be difficult to operate.

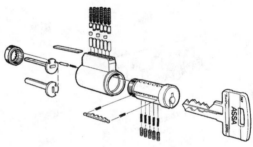

65631 Yale

1. Install cylinder plug in housing as shown in diagram.

2. Install cylinder pug retainer "C" clip from above and ensure it is well seated.

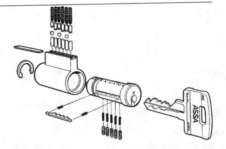

ASSA TWIN 6000

EQUIPMENT AND SUPPLIES
Pin kit, tools, and samples

Order No. PK-2

Pin Kit includes 100 pieces of size each bottom, master and drive pins, 500 tumbler springs, 50 sidebars, 100 each side pins and sidebar springs, 100 Phillips cam screws, and 25 mortise/rim cover strips.

Order No. 86 90 18

ASSA Twin cutter wheel for HPC 1200 CM 90°/.032"

Order No. 96 90 20

ASSA Twin code card for HPC 100 CM

Order No. 90 70 09

Clear Deadbolt Mounting Ring.

Order No. 90 03 05

Deadbolt Sample Mount (base only).

Order No. 87 26 30

ASSA Twin Mortise Cutaway.

ASSA High Security Locks

3475 14th Avenue
Markham, ON L3R 0H4
Canada
Phone: 905 940-2040
Fax: 905 940-3242

An ASSA ABLOY Group Company

6

Rekeying
Kwikset Locks

If you can rekey Kwikset locks, you can rekey almost any standard pin tumbler lock. In this chapter, you learn all you need to know about how to rekey a Kwikset lock.

The rest of this chapter is a reprint of the Kwikset Rekeying Manual, courtesy of Kwikset Corporation. If you learn to rekey Kwikset locks, you should be able to rekey most standard pin tumbler locks.

Rekeying Is A Great Customer Service That's Easy To Learn.

When customers buy a new entry lock-set, deadbolt or handleset, they also want the convenience of having one key operate all the locks in their home. Most customers are not aware that rekeying servic-es exist. They go through life adding more and more keys to an already overcrowded key ring. Locksmiths pro-vide valuable rekeying services and, for extra customer conven-ience, so are more and more hardware stores. That's where you come in. You can help make your account more profitable and more convenient for their customers.

Let's say you sell a new Kwikset Security, Maximum Security, UltraMax Security or Society Brass deadbolt. Ask your customer if he or she also has a entry lockset. If so, would they like you to make their existing key work with the new deadbolt? It will only take a few minutes and cost a few dollars. Most customers will consider that a real convenience and you'll have made an extra sale.

Naturally, it works in reverse, and it works if the customer buys both a deadbolt and an entry lockset or handleset. You ask for their existing key, follow the procedures you'll learn in this manual and you have a happier customer. If they have a competitors brand, the Kwikset Keying Station may allow you to exchange UltraMax cylinders for their existing brand.

Before we go into the mechanics of how one key can do it all, first let's define some terms, then take a quick look at how locks work.

Terminology: When keying two locks alike, you are actually changing the pin-tumblers in one of the locks so that it will have the same "cut combination" as the other. Therefore, "repinning" is a more accurate word to use than "rekey-ing" But that's what it has been called for generations and we're not about the change that. "Keying alike" is just another term for rekeying.

UltraMax's Superior Engineering Makes Rekeying Even Easier.

If your customer has chosen the UltraMax brand, your job is even easier (in most cases).

When your customers purchase an UltraMax knob, dead-bolt, lever, or handleset, they have selected a superior product which provides them with the extra security and peace of mind they need. A single key for all their locks in their home adds security as well as convenience. That feeling of security is what your customers are looking for. All you need is the key that they would like to use for all of their locks.

UltraMax & The Society Brass Collection's six-pin key.

Some Kwikset locks, and most other brands as well, are made with a six-pin cylinder so there are more key cut combinations, which make it more difficult to pick and less likely that any two peo-ple will have matching keys. It does not mean that it is more difficult to rekey. In fact, in some circumstances, it's easier. And there is no problem keying a six-pin lock to match a five-pin lock.

The Society Brass Collection uses the Solid Brass six-pin cylinder, eliminating the need of lubricant, which makes it easier to re-key, as well as eliminating the chance of the pins collecting dust and sand.

UltraMax's removable cylinder: UltraMax's knobs have a special cylinder that is removable in seconds with the control key. This will save you time and be convenient for your customers.

In the pages that follow, you will learn how to rekey a Kwikset entrance lockset, a key-in-lever, a single-piece entrance handleset and a security deadbolt. While most of these techniques are the same, there are some subtle differences in how you take the locks apart and put them back together or change out the cylinders.

You'll also learn how to gauge the cuts of a key so you will be able to read it's "combination" and select the prop-er size pins. Let's get underway.

Contents

Glossary of Terms

bitting . . . . .key cuts that form the combination.

blade . . . . .the portion of a key and/or milling.

bow . . . . .the portion of the key which serves as a grip or handle.

bow stop . . . . .a type of stop located near the key bow.

chamber . . . . .any cavity in a cylinder plug and/or shell which houses the tumblers.

combinate . . . . .the pinning of the actual cuts on the key.

Control Key . . . . . . .available for most TITAN products, allows removal of cylinder without dismantling lock.

cylinder assembly . . .holds the plug, pins, springs and cover.

cylinder guard . . . . .heavy duty cover for cylinder housing.

cylinder housing . . . .contains the cylinder assembly.

cylinder removing tool .for removing and replacing plug retainer.

gauge . . . . .the act of determining the bitting of the key

handing . . . . .the orientation of the knob, handle, or lever with respect to a left or right sided door (see illustration on page 20).

lever catch . . . . . . .spring mounted piece, holds lever in place

locking bar . . . . . . .prevents removal of UltraMax cylinder without Control Key use or retainer removal.

mechanism module . .keyed entry lockset assembly for new keyed handlesets.

PK holes . . . . .Protecto Key, allows new homes to be keyed to a builder's key, and once homeowner uses their key, builder is locked out.

plug . . . . .holds the bottom pins and keyway.

plug follower . . . . . .used to push the plug out and hold top pins and springs in place while rekeying plug.

plug retainer . . . . . ."C" style clip, holds plug in cylinder housing.

shear line . . . . .when the bottom pins are correct, a shear line is created when the proper key is inserted allowing rotation of plug.

shield . . . . .for additional security on exterior knobs.

sleeve . . . . .assembly holding cylinder and lever.

spindle . . . . .two types; a round and half-round spindle.

spring cover . . . . . . .snap on cover to hold pin springs in place.

spring housing . . . .holds return spring for lever or knob in place.

tailpiece . . . . .spindle like extension for deadbolts.

thumbpiece pin . . . .holds the handlesets thumbpiece in place.

How a Lock Works

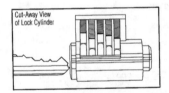

Cut-Away View
of Lock Cylinder

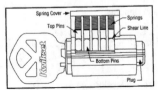

Spring Cover
Top Pins
Springs
Shear Line
Bottom Pins
Plug

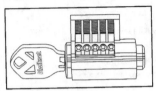

5-Pin System

Kwikset locks operate by matching the cuts on the keys with the bottom pin tumblers inside the cylinder plug.

There are two sets of five pins in each lock, top and bottom, and a set of springs. The top pins are all the same size and are flat on both ends. You do not, at least right now, want to deal with top pins or springs.

You only want to deal with bottom pins, which are of different lengths (in .023" increments) and are tapered on both ends.

For the lock to work, the cuts of the key must enable all five bottom pins to be flush with the cylinder plug. This is called the shear line.

In the top photo, there is no shear line because some bottom pins are out of place. That key won't operate this lock.

Put in the correct key (middle photo) and all the pins line up to form the shear line and the key will operate the lock (bottom).

When you rekey a lock, you simply replace the bottom pins according to the cut combination of the key you want to use. And you can do all this with a few very simple tools.

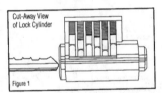

Cut-Away View
of Lock Cylinder

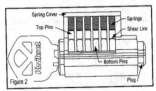

Figure 1

Spring Cover
Top Pins
Springs
Shear Line
Bottom Pins
Plug

Figure 2

6-Pin System

UltraMax Security locks operate by matching the cuts on the keys with the bottom pin tumblers inside the cylinder plug.

There are two sets of six pins in each lock, top and bottom, and a set of springs. The top pins are all the same size and are flat on both ends.

You only want to deal with bottom pins, which are of different lengths (in .023" increments) and are chamfered on both ends.

For the lock to work, the cuts of the key must enable all six bottom pins to be flush with the cylinder plug. This area is called the shear line.

In the top photo (Figure 1), with no key or the incorrect key, some bottom pins and some top pins block the shear line. That key won't operate this lock.

In the middle photo (Figure 2), with the correct key, all the pins line up at the shear line. The key will turn and operate the lock (Figure 3).

When you rekey a lock, you simply replace the bottom pins according to the "bitting" (cut combination) of the key you want to use.

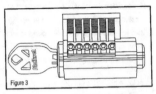

Figure 3

2

Tools to Rekey

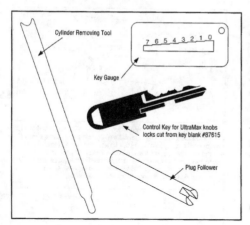

Cylinder Removing Tool

Key Gauge

7 6 5 4 3 2 1 0

Control Key for UltraMax knobs
locks cut from key blank #87615

Plug Follower

Inside a Kwikset/Society Brass Collection Rekeying Kit are the different bottom pin sizes you need to rekey a lock, a Key Gauge for reading the cuts on a key, a Cylinder Removing Tool (affectionately known as a "pickle fork"), and a plug follower. This is a very simple device which keeps lock parts from scattering across the room when you remove the plug (which houses the pins, and into which the key fits) from the cylinder.

There are also extra top pins, springs and other parts in the kit, but you do not need to be concerned with those now.

(Shown: Keying Kit No. 272 contents)

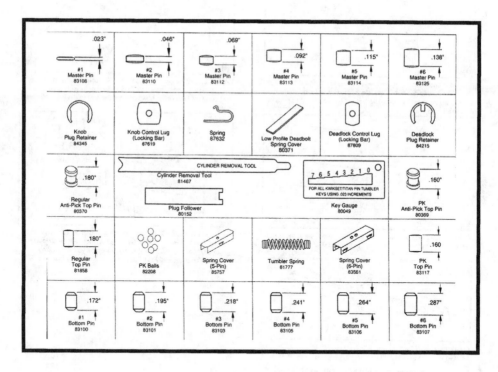

.023" #1 Master Pin 83108	.046" #2 Master Pin 83110	.069" #3 Master Pin 83112	.092" #4 Master Pin 83113	.115" #5 Master Pin 83114	.138" #6 Master Pin 83125
Knob Plug Retainer 84345	Knob Control Lug (Locking Bar) 87619	Spring 87632	Low Profile Deadbolt Spring Cover 80371	Deadlock Control Lug (Locking Bar) 87809	Deadlock Plug Retainer 84215
.180" Regular Anti-Pick Top Pin 80370	CYLINDER REMOVAL TOOL / Cylinder Removal Tool 81467 / Plug Follower 80152			7 6 5 4 3 2 1 0 FOR ALL KWIKSET/TITAN PIN TUMBLER KEYS USING .023 INCREMENTS Key Gauge 80049	.160" PK Anti-Pick Top Pin 80369
.180" Regular Top Pin 81858	PK Balls 82208	Spring Cover (5-Pin) 85757	Tumbler Spring 81777	Spring Cover (6-Pin) 83561	.160 PK Top Pin 83117
.172" #1 Bottom Pin 83100	.195" #2 Bottom Pin 83101	.218" #3 Bottom Pin 83103	.241" #4 Bottom Pin 83105	.264" #5 Bottom Pin 83105	.287" #6 Bottom Pin 83107

How to Gauge a Kwikset Key

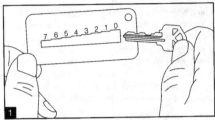

Before you can rekey a lock, you have to know what pins to use. For obvious security reasons, Kwikset doesn't print key-cut combinations on the packaging. We use this Key Gauge to find the key-cut combinations. Before disassembling the lock, measure the cuts and write down the numbers.

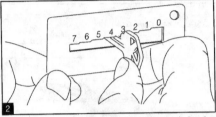

Hold the NEW key (the one with which you want this lock to be keyed alike) and slide it into the gauge. ALWAYS gauge a key from the bow (the bow being the part you hold in your fingers) end out. To measure, position the flat portion of the first cut even with the "0" position of the gauge, slide the key toward the narrowing end of the gauge until the key stops at the correct "step". This will always be between two numbers and the cut number is the one to the RIGHT of the key. Here it is a 3.

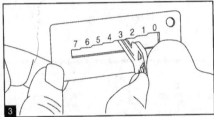

Position the next cut and move the key down the gauge until it stops. The second cut of this key is a 2. You can slide the gauge or the key, the result is the same. The first cut comes after the shoulder next to the bow.

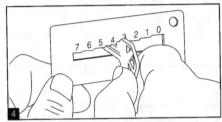

Do the same for the third cut which is a 4. Remember, always read the number to the right of the key. Also, remember to read the cuts from the bow of the key out (the bow being the part you hold with your fingers).

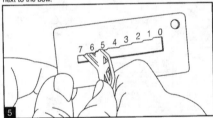

It's also a good idea to double-check that you are reading the correct cut each time. In this case, cut number four is a 6.

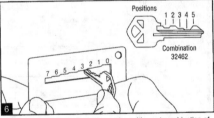

Finally, the last cut: a 2. So we have a key with a cut combination of 3-2-4-6-2 (if this is a UltraMax Security key, there will be six cuts to gauge). Of course, you've written that down as you went along. When the time comes, you'll know exactly which pins to select.

Locate the special "control key" which is cut specifically for the combination of your lock, but with a notch on the bottom of the key blade.

NOTE: If not available, cut a blank "control key" to match existing key.

4

Key Duplication

Instructions for duplicating cuts from a five pin Kwikset key to a six pin Kwikset or Society Brass key. This permits Kwikset and Society Brass to be keyed alike.

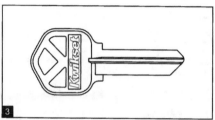

Measure the Kwikset key and write down the key cut combination. These will be the depths required for positions 2 through 6 on the 6 pin key. Save the results for re-pinning the cylinder.

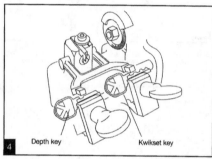

Kwikset key   Key blank

The key duplicator must be of the type that stops against the keys top shoulder. Insert both the Kwikset key and the key blank. Positions 2 through 6 are then cut. Position 1 will be bypassed.

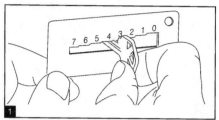

Depth keys are used to duplicate the remaining cut in position 1 of the key. Select depth at random from the 3 keys supplied. Only position 1 has been cut, to prevent contact with any other cut. The cylinder will be pinned to this key cut.

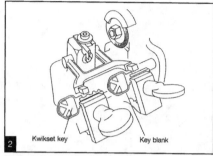

Depth key   Kwikset key

Insert both the depth key and the previously cut key in the key duplicator. Now position 1 of the key is cut to match the depth key. Additional keys may now be duplicated in the usual manner.

Masterkeying

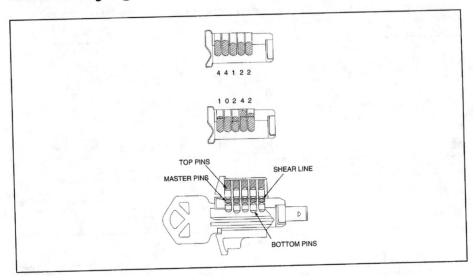

4 4 1 2 2

1 0 2 4 2

TOP PINS

MASTER PINS

SHEAR LINE

BOTTOM PINS

Masterkeying
Masterkeying is making one lock work with two keys. Take two keys - one will be the master key which will be able to open all locks in the complex and the other will be the random key or the tenant's key, which only opens one lock.

1   Read both keys. (see Step 1, page 5)

Master Key	5	4	3	2	2
Random Key	4	4	1	6	4

2   Circle the smallest number
    of each cut.

Master Key	5	4	3	(2)	(2)
Random Key	(4)	(4)	(1)	6	4

The circled numbers are your bottom pins required.
Drop in the bottom pins without the key in the cylinder.
EXAMPLE - 4 4 1 2 2

3   Take the difference of each cut.

Master Key	5	4	3	2	2
Random Key	4	4	1	6	4
	1	0	2	4	2

The difference (10242) will be the master pins required.
Drop in these master pins on top of the bottom pins
without the key in the cylinder. A 0 means that a
master pin is not put in that hole.

4   Put plug back into the cylinder assembly without
    key in the plug.

6

Keying Master Cylinders

Model 740 cylinder is illustrated. Other cylinders are keyed in same manner.

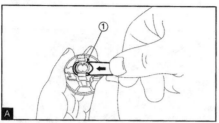

Remove plug clip
Push plug clip (1) off with screwdriver or cylinder removing tool, using a side-to-side motion.

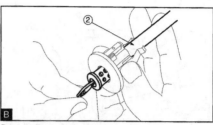

Remove plug
Insert key and turn plug 1/4 turn clockwise. Push plug out of cylinder with plug following tool (2) and leave in cylinder. the plug following tool will keep the top tumbler pins and springs in place.

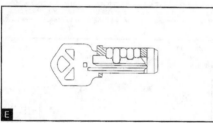

Drop pins out of plug.

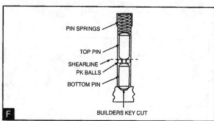

Key measurement
Measure cuts of key or keys as illustrated, and write down key cut combinations.

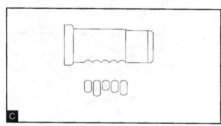

For random keying
Random keying uses only one key. For multiple keys, refer to page 6 (Masterkeying). Insert key into plug. Drop bottom pins into proper holes with pin numbers that match measured depths. Tops of pins should be flush with outside diameter of plug.

Protecto Keying
Larger kits are capable of Protecto Keying. This is used for new construction and similar to Masterkeying. Use 3 balls in place of a #2 master Pin. Builders key will have a cut two increments deeper than homeowners random key in selected chamber.

Rekeying Knobs

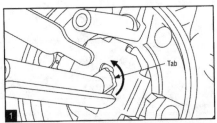

Be sure to remove key from cylinder before starting. In order to remove the round spindle, align the tab so it is perpendicular to the bottom of the half-round, as shown at right.

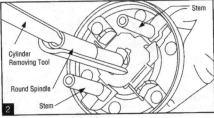

Insert the pointed end of the Cylinder Removing Tool into the end of the spindle and turn until the tab is lined up with the stems.

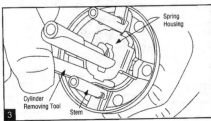

Insert small end of the Cylinder removing Tool at an angle against stem, sliding under edge of spring housing. Push in firmly on retainer.

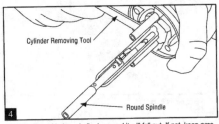

Hold the lock with the spindle down and it will fall out. If not, keep pressing with the tool and use your thumb to slide the round spindle out. **If the round spindle rotates, you will need to repeat steps 1 - 4.**

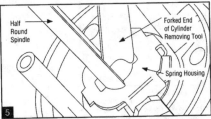

To remove the cylinder, place the forked end of the tool down the half-round spindle through spring housing. You'll see a hole there, and you have to fit the forked end of the Cylinder Removing Tool under the top of that hole and pry it open.

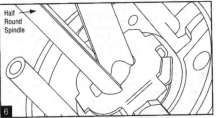

Once you have "opened" the spring housing, slide the tool all the way in until it stops. Be firm, but not forceful. Keep the Cylinder Removing Tool perpendicular to the half-round spindle at all times.

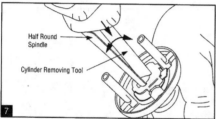

With the Cylinder Removing Tool all the way in, put your other hand over the outside of the knob, covering the cylinder and plug.

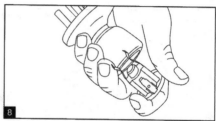

The cylinder will come right out in your hand. For most, this is the most difficult part of rekeying. It takes practice to get the tool in under the spring housing, and getting just the right motion and pressure to pop the cylinder loose.

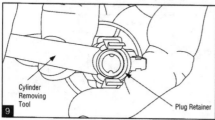

Use the forked end of the Cylinder Removing Tool to disengage plug retainer.

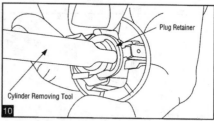

Then use the small end to free it completely. Use care not to deform plug retainer.

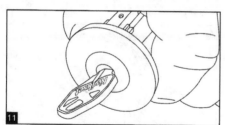

Slide the EXISTING key for this lock into the keyway and turn the key about 45° to the right or left. Keep key at this angle for next step. **Be sure not to let key cylinder slide out.**

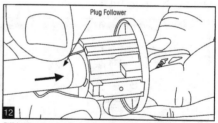

Still holding the plug firmly in the cylinder, use the **flat end** of the plug follower (the notched end is for a different kind of cylinder) to PUSH the plug out gently and evenly from the rear. Push it steadily, do not pull the plug.

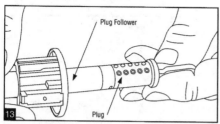

Keep pushing until the tool sticks out at least an inch. **Leave it in place until you reassemble.** Carefully remove the plug, holding it by the key and don't turn it.

Now you can see the bottom pins in the plug. Turn it over to dump them out and discard them. It's not worth the time and possible sorting errors to restock them.

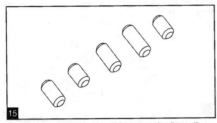

Checking the key bitting (combination) you gauged earlier, you'll select the pins which match those cuts from the appropriate compartments in the Rekeying Kit. Example: 3-2-4-6-2. A No. 3 bottom pin to a No. 3 key cut depth, a No. 2 to a No. 2, etc.

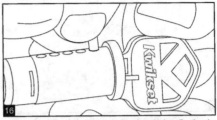

Put the new key into the plug and, working from the bow (front) end of the plug, drop the first pin into its chamber – here a No. 3. Some people find using their fingers easier, others prefer tweezers. Either way, it takes some patience to deal with these tiny pins.

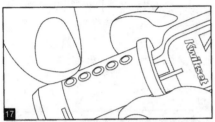

Install pins one by one, from the front (bow) to the back (tip) and be sure they are all flush with the top surface of the plug. A pin which is too long will prevent plug insertion with the key installed. A pin which is too short will cause the lock to malfunction.

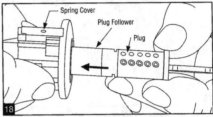

Now all you have to do is reassemble the plug into the cylinder housing in exactly the reverse order. Start by placing the end of the plug against the end of the plug follower. Again, turn the key so it is about 45° to the right or left of the spring cover. Then slowly and smoothly PUSH the plug follower back into the cylinder with the plug.

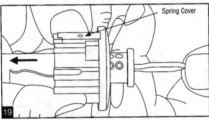

When it's all the way in the cylinder, turn the key so it is in line with the spring cover on top of the cylinder. You're done with the plug follower, but hold the plug and cylinder firmly together.

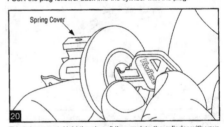

This is important. Hold the plug all the way into the cylinder with your thumb. Making sure the key is still in line with the pin cover, pull out the key. Keep pressure on the plug until the key is out.

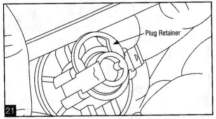

Put the plug retainer back in from the side of the spring cylinder as shown and use the Cylinder Removing Tool to snap it back into place. Be sure the inside edges of the retainer line up with grooves in the plug.

Insert the key and turn it back to the vertical position, in line with the spring cover on top of the cylinder.

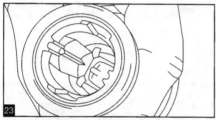

Observe the wide channels top and bottom inside the knob cavity and the spring cover on top of the cylinder. Correct orientation of the knob, will depend of the handing of the doorway. With the curved side of the half-round toward your body, align the pin cover with the top channel in the knob cavity.

Keeping the spring cover aligned with the top channel, push the cylinder all the way into the knob until you hear the spring retainers snap into place.

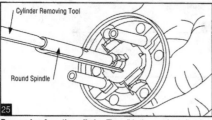

Remove key from the cylinder. Then slide the round spindle tab-end first and tab up, into the half-round spindle and push it in until it clicks.

Reinsert the key into the cylinder. Hold assembly by the stems and work the key back and forth a few times to make sure everything is in proper order.

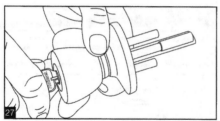

You have rekeyed a Kwikset lockset!

Rekeying Laurel Eggknob

Step-by-step instructions for:
Kwikset
UltraMax
SECURITY

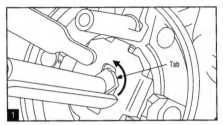

Tab

Be sure to remove key from cylinder before starting. In order to remove the round spindle, align the tab so it is perpendicular to the bottom of the half-round, as shown at right.

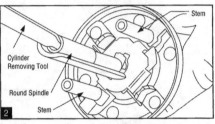

Stem
Cylinder Removing Tool
Round Spindle
Stem

Insert the pointed end of the Cylinder Removing Tool into the end of the spindle and turn until the tab is lined up with the stems.

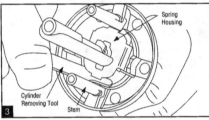

Spring Housing
Cylinder Removing Tool
Stem

Insert small end of tool at an angle against stem, sliding under edge of spring housing. Push in firmly on retainer.

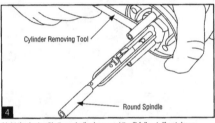

Cylinder Removing Tool
Round Spindle

Hold the lock with the spindle down and it will fall out. If not, keep pressing with the tool and use your thumb to slide the round spindle out.
If the round spindle rotates, you will need to repeat steps 1 - 4.

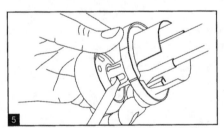

Depress the lever catch and pull keyed portion from the chassis.

Pull off knob cover (shank).

7

Turn knob over and the cylinder will fall out into your hands. At this point, you are able to re-key the cylinder.

8

After re-keying the cylinder, return it to its original position, while holding the keyed portion toward the floor.

9

Replace the knob cover. (note: it goes on either way) Make sure that it has a snug fit to avoid any possible "knob wobble".

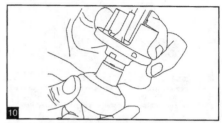

10

Attach keyed portion to the chassis. Make sure that you hear the lever catch engage. Pull on the keyed portion to the chassis. Make sure that you hear the lever catch engage. Pull on the keyed portion to be sure that it is properly re-assembled.

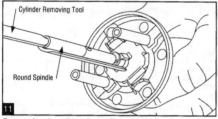

Cylinder Removing Tool

Round Spindle

11

Remove key from the cylinder. Then slide the round spindle tab-end first and tab up, into the half-round spindle and push it in until it clicks.

12

Reinsert the key into the cylinder. Hold assembly by the stems and work the key back and forth a few times to make sure everything is in proper order.

13

Rekeying Knobs
with a control key

If no control key exists, use the customer's existing key to cut one. When making the new key, also make a new control key with the same bitting (combination) as the new key.

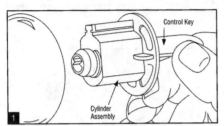

Insert control key, turn 90° (1/4 turn) counterclockwise and pull out core. Remove control key from core. Failure to do this could allow plug to come out of the cylinder shell dropping pins and springs.

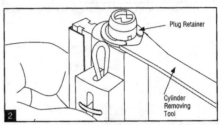

Fit forked end of Cylinder Removing Tool into open end of plug retainer and push it out (use screwdriver if tool is not available).

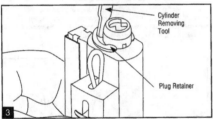

Move the plug retainer out the rest of the way with the other end of the Cylinder Removing Tool (or small screwdriver).
Use care not to deform retainer.

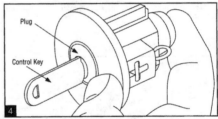

Reinsert old control key. Use thumb to hold key or plug in place. (With the plug retainer off, the plug could slide out, dropping pins and springs.)

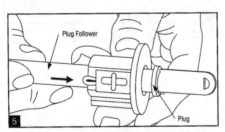

Rotate plug slightly clockwise or counter-clockwise. Using the follower, push the plug out from the back. Make sure the follower stays tight against the plug or the top pins and springs will fall out. (See page 19 if this happens.) Make sure the plug follower sticks out at least one inch. **Leave it in place until you reassemble.**

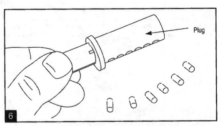

Drop out the old pins, remove the old control key and insert the new control key. (We will assume you have already gauged the new control key.)

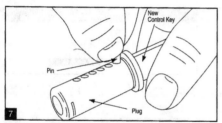

7

Begin inserting the new bottom pins from the appropriate compartments in the rekeying kit according to the bitting of the new key. You may handle the bottom pins with tweezers or your fingertips.

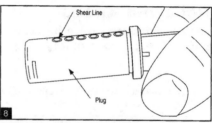

8

If all the bottom pins are the proper size for the key, they will create the shear line. **Check this carefully.** A pin which is too long will prevent insertion of the plug while the key is inserted. A pin which is too short will allow the cylinder to lock up with no key operation.

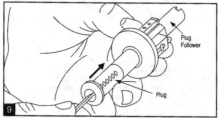

9

Start reassembly in the same manner you began. Use the plug to push the follower back through the cylinder. Hold the plug firmly into the cylinder with your thumb and turn the control key straight up and remove it.

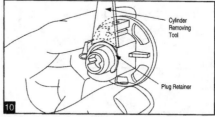

10

Use the cylinder removal tool to reseat the plug retainer. Test cylinder with key to make sure it turns properly.

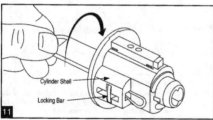

11

Insert and rotate control key 90° (¹/₄ turn) counterclockwise. Locking bar must be flush with the surface of the shell for insertion into knob.

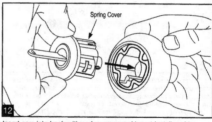

12

Insert core into knob with spring cover up (do not install upside down) if lockset is not on door.
NOTE: Correct orientation depends on the hand of the door where the lock is to be installed.

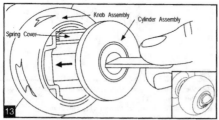

13

Insert core into knob. Make sure that the spring cover is in the up position.
Note: Interior turnbutton must be in the unlocked position.

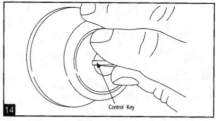

14

Rotate control key 90° (¹/₄ turn) clockwise to vertical position and remove. Check with standard operating key for proper operation.

Rekeying Knobs
with a standard operating key but without a control key

If you can cut a control key to match the existing key, the process is much easier.

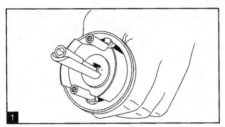

Remove lockset from door.

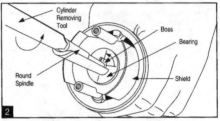

Pull the shield off far enough to allow access to the mechanism at the base of the spindle. Rotate the round spindle with the Cylinder Removing Tool (or small screwdriver) to line up the boss of the spindle with the slot of the plastic bearing.

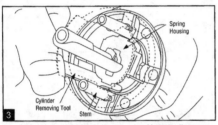

Insert the same end of the cylinder removal tool between the screw stem and the detent slide. Apply slight force by pressing against detent slide to remove spring tension while holding spindle downward allowing spindle to drop out. If necessary, use free finger to assist spindle removal. If spindle rotates, repeat step (2).

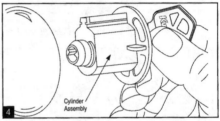

Turn key 180° (½ turn) counterclockwise and pull out cylinder.

CAUTION: Turn the key back and remove it. Failure to do this may allow the plug to come out, dropping pins and springs if plug retainer is removed first.

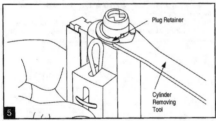

Fit forked end of Cylinder Removing Tool into open end of plug retainer and push it out (use screwdriver if tool is not available).

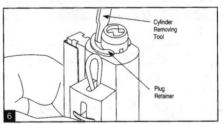

Remove the retainer completely using the other end of the tool (or small screwdriver).

NOTE: Use care not to deform plug retainer.

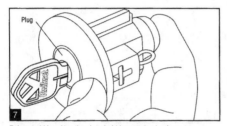

7

Reinsert old key. Use thumb to hold key or plug in place.

CAUTION: With the plug retainer off, the plug could slide out, dropping the pins and springs.

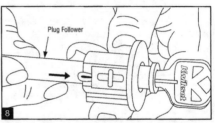

8

Use the plug follower to push the plug out from the back. Make sure it stays tight against the back of the plug or the top pins and springs will fall out. (See page 19 if this happens.)

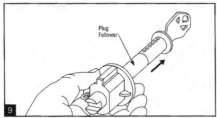

9

Make sure the follower sticks out at least one inch.
Leave it in place until you reassemble.

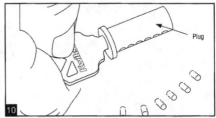

10

Drop out the old pins, remove the old key and insert the new key. (We will assume you have already gauged the new key.)

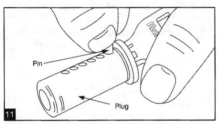

11

Begin inserting the new pins from the appropriate compartments in the rekeying kit according to the bitting (combination) of the new key. You may handle the bottom pins with tweezers or your fingertips.

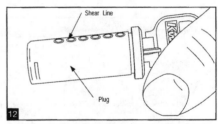

12

If all the bottom pins are the proper size for the key, they will create the shear line. **Check this carefully.** A pin which is too long will prevent insertion of the plug while the key is inserted. A pin which is too short will allow the cylinder to lock up with no key operation!

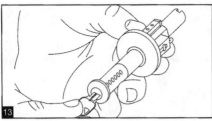

13

Start reassembly in the same manner you began. Use the plug to push the follower back through the cylinder.

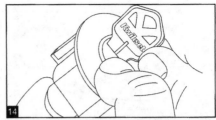

14

Holding the plug firmly in the cylinder with your thumb, turn the key straight up and remove it.

CAUTION: Do not allow the plug to come out as you remove the key!

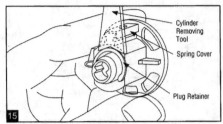

15 Use the cylinder removing tool to reseat the plug retainer. Test cylinder with key to make sure it turns properly.

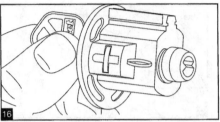

16 Insert key and turn 180° (½ turn). The cylinder can be installed into the knob two ways. To be sure it will be right side up when the lockset is installed, you must determine the hand of the door before installing the cylinder.

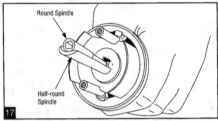

17 Notice C-shaped outer spindle protruding behind knob. Orientation of the lockset on the door is determined by this piece.

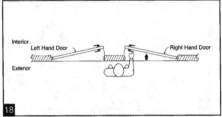

18 Picture the door from the keyed side (normally the outside) and determine whether the hinges are on the right or left. The knob assembly will be right side up when the open side of the C-shaped spindle faces the hinges and the smooth curved side faces the free edge of the door.

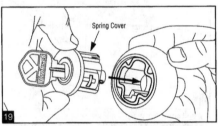

19 Hold the knob assembly right side up for its eventual installation. Align the cylinder spring cover with the top of the knob and install the cylinder.

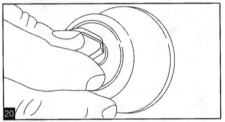

20 Turn the key 180° (½ turn) and remove key.

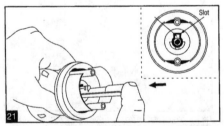

21 Replace the round inner spindle by lining up boss with slot. Push spindle until it snaps into place. Reseat shield and plastic bearing.

18

Kwikset UltraMax SECURITY

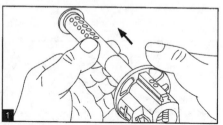

Pull out the plug and plug follower allowing all pins and springs to fall out of the cylinder.

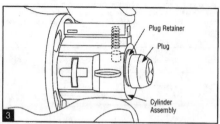

After plug is rekeyed per the instructions, insert into cylinder shell. Install plug retainer. Replace all top pins and springs (they are all the same). Carefully push spring cover on. If you have removed the locking bar and spring, replace them at this time. Place locking bar in slot. Rotate key until locking bar is flush with cylinder shell. Secure with locking bar spring. Insert the key and test for proper operation.

Note: The plug follower is not required while reassembling.

Rekeying Knobs
What to do if the top pins and springs drop out during rekeying

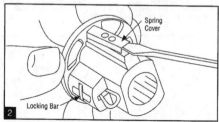

Carefully pry off spring cover with small screwdriver. Removal of the locking bar is not necessary. When you reach the point to reinsert the plug into the cylinder housing, depress the locking bar from the inside of the cylinder housing with a small screwdriver while inserting the plug.

Rekeying Levers

DOOR HANDING

Interior
Left Hand Door   Right Hand Door
Exterior

To determine the handing of a door, face the OUTSIDE of the door. If the hinge is on the left side of the door, the door is left hand (LH). If the hinge is on the right side of the door, the door is right hand (RH).

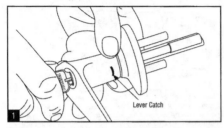

Lever Catch

Insert the key and unlock the leverset. The lever cannot be removed if the leverset is in the locked position. Remove the key.

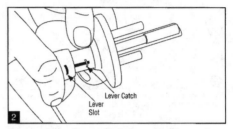

Lever Catch
Lever
Slot

Using a small screwdriver, press in the lever catch and remove the lever from the assembly (wiggling the lever a little will aid with its removal). The cylinder assembly will be in the sleeve or the removed lever. Remove it.

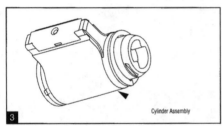

Cylinder Assembly

The cylinder assembly is now ready to be rekeyed. Rekeying this cylinder is the same as steps 7 - 11 on page 17.

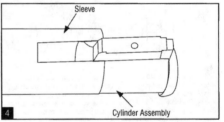

Sleeve

Cylinder Assembly

After rekeying the cylinder assembly insert the cylinder into the outside sleeve as shown. Some wiggling may be required for the cylinder assembly to seat fully in the sleeve.

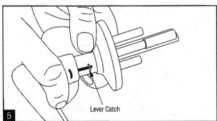

Lever Catch

Slide the lever over the sleeve and cylinder assembly.

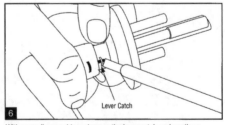

Lever Catch

With a small screwdriver, depress the lever catch and continue pushing and slightly rotating the lever onto the sleeve until the catch locks the lever in place. The leverset is now rekeyed.

Step-by-step instructions for:

Rekeying UltraMax Levers
For Commercial Series Levers

Control key for UltraMax Commercial #81804
(when cutting the above control key, do not remove control key tip)

For when a control key is not available, but the homeowners key is.

If you can cut a control key to match the existing key, continue with steps 5–13 in UltraMax Knobs on page 14.

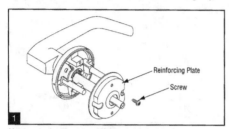

Reinforcing Plate

Screw

1

Unscrew and remove the reinforcing plate (1 screw).

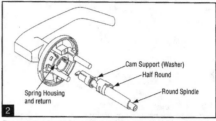

Cam Support (Washer)

Half Round

Round Spindle

Spring Housing
and return

2

Remove round spindle assembly (round spindle, half-round, and washer). Keep assembly together.
CAUTION: Do not disturb spring housing and return.

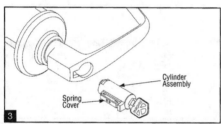

Cylinder
Assembly

Spring
Cover

3

Insert key and turn 180° to the position shown. Pull out the cylinder assembly (if homeowners key is not available, cylinder must be picked). Continue with steps 5–13 in "UltraMax Knobs on page 14 - 15.

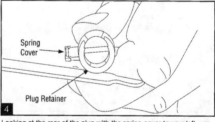

Spring
Cover

Plug Retainer

4

Looking at the rear of the plug with the spring cover to your left, use the cylinder removing tool to re-seat the plug retainer from the bottom as shown. Insert key, and turn180° and replace cylinder assembly into the lever, turn key 180° and remove. Spring return and housing must be seated properly for the cylinder to lock in place.

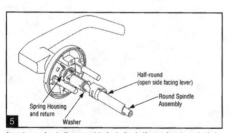

Half-round
(open side facing lever)

Round Spindle
Assembly

Spring Housing
and return

Washer

5

Insert round spindle assembly (spindle, half-round and washer), in opening by aligning the paddle, at the end of the spindle assembly, to the matching slot in cylinder. Once assembled, half-round opening faces lever.

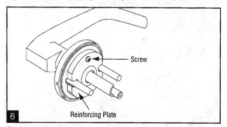

Screw

Reinforcing Plate

6

Replace reinforcing plate and screw down. If the plate is not flush, the spindle assembly is not in proper position.

Step-by-step instructions for:
Kwikset UltraMax SECURITY

Rekeying UltraMax Levers
For Commercial Series Levers

Control key for UltraMax Commercial #81804
(when cutting the above control key, do not remove control key tip)

For when a control key is not available, but the homeowners key is.

If you can cut a control key to match the existing key, continue with steps 5–13 in UltraMax Knobs on page 14.

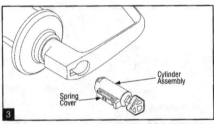

1

Unscrew and remove the reinforcing plate (1 screw).

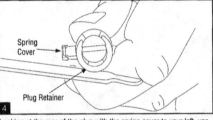

2

Remove round spindle assembly (round spindle, half-round, and washer). Keep assembly together.
CAUTION: Do not disturb spring housing and return.

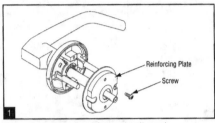

3

Insert key and turn 180° to the position shown. Pull out the cylinder assembly (if homeowners key is not available, cylinder must be picked). Continue with steps 5–13 in "UltraMax Knobs on page 14 - 15.

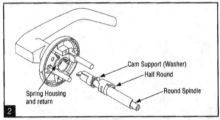

4

Looking at the rear of the plug with the spring cover to your left, use the cylinder removing tool to re-seat the plug retainer from the bottom as shown. Insert key, and turn180° and replace cylinder assembly into the lever, turn key 180° and remove. Spring return and housing must be seated properly for the cylinder to lock in place.

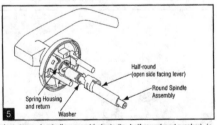

5

Insert round spindle assembly (spindle, half-round and washer), in opening by aligning the paddle, at the end of the spindle assembly, to the matching slot in cylinder. Once assembled, half-round opening faces lever.

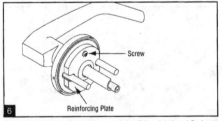

6

Replace reinforcing plate and screw down. If the plate is not flush, the spindle assembly is not in proper position.

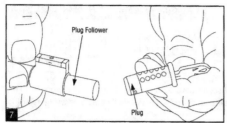

Insert the new key. Dump and discard the old pins (If you didn't gauge the key before starting, do it now and lay out the pins in order — from the bow to tip). Leave new key in place.

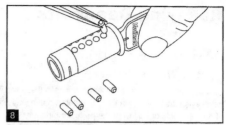

Begin dropping in the appropriate pins (this is a 6-5-4-4-5). If all the bottom pins are the proper size for the key, they will be at the shear line - even with the surface of the plug. **Check this carefully.** A pin which is too long will prevent insertion of the plug while the key is inserted. A pin which is too short will allow the cylinder to lock up with no key operation!

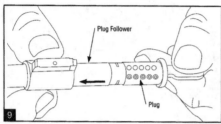

Holding the key 45° right or left of vertical, use the plug to push the follower back through the cylinder.

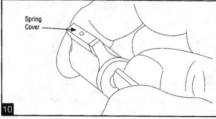

Turn the key so it's in line with the spring cover on top of the cylinder. Hold your thumb against the plug face and pull the key out carefully. Do not allow the plug to slide out.

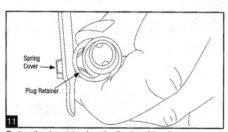

Replace the plug retainer from the direction of the pin cover. Remember to align the open end with the groove in the plug. **CHECK CAREFULLY-If plug retainer is replaced incorrectly, the lever may not lock onto sleeve.**

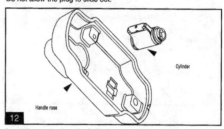

Insert the newly pinned cylinder back into the handle rose.

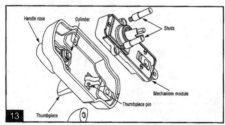

Place the thumbpiece into the handle rose slot as shown and slide the thumbpiece pin through the hole in the back of the thumbpiece. The mechanism module keeps the pin in place. Attach the mechanism module with the 2 stud screws.

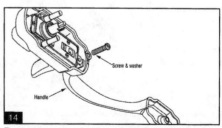

The remaining screw and washer assembles the bottom hole in the mechanism module with the handle.

Rekeying Deadbolts

Step-by-step instructions for:
Kwikset Security   Kwikset Maximum SECURITY

A lot of people are buying Kwikset deadbolts in these security-conscious days. And when they do, you're the one to suggest keying them alike with their entry locksets. In rekeying a deadbolt, the principle is exactly the same as the rekeying of an entry lock, but a few of the parts are different.

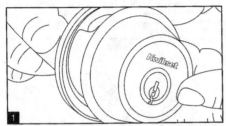

With single cylinder deadbolts, you'll only be dealing with the exterior side of the lock. With double cylinder locks, both sides must be rekeyed.

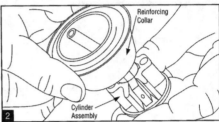

Turn the cylinder face down and remove the reinforcing collars.

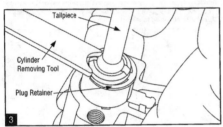

To remove the tailpiece, fit the forked end of the Cylinder Removing Tool into the open end of the plug retainer and push it out.

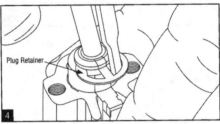

Move the plug retainer out the rest of the way with the other end. **Use care not to deform retainer.**

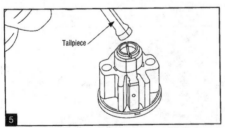

Remove the tailpiece and proceed the same as you did with the knob lock.

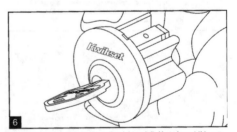

Insert the old key, turn it 45° to the right or left. Keep key at this angle for next step.

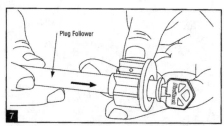

7

Use the flat end of the plug follower to push the plug slowly and evenly out from the back. Make sure the follower sticks out at least an inch. **Leave it in place until you reassemble.**

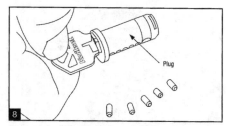

8

Dump and discard the old pins. Remove old key and insert new key. (We assume you have already gauged the new key.)

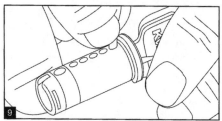

9

Install the new pins according to the bitting (combination) of the new key. As mentioned earlier, whether you handle the pins with tweezers or your fingertips is your choice.

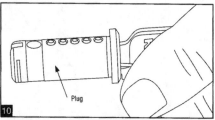

10

If all the bottom pins are the proper size for the key, they will be at the shear line -- even with the top surface of the plug. **Check this carefully.** A pin which is too long will prevent insertion of the plug while the key is inserted. A pin which is too short will allow the cylinder to lock up with no key operation!

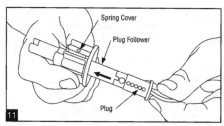

11

Reassemble the same way. Turn the key 45° to the right or left of the spring cover. Use the plug to push the follower back through the cylinder gently and evenly.

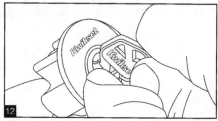

12

Holding the plug firmly in the cylinder with your thumb, turn the key straight up and remove it carefully.

CAUTION: Do not allow plug to come out as you remove the key!

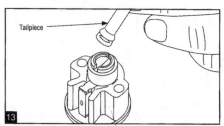

13

Install the tailpiece. If servicing a double cylinder model, be sure each cylinder has the right tailpiece. Align the notches in both the plug and tailpiece.

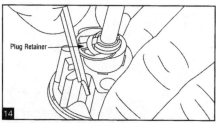

14

Use the Cylinder Removing Tool to reseat the retainer. Insert the key and test for proper operation.

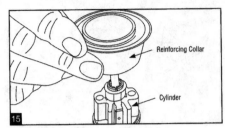

Put the reinforcing collar and cover back on.

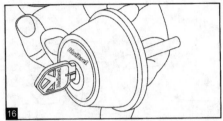

Your deadbolt rekeying is complete. If your customer has purchased a Kwikset double cylinder model, simply rekey the other cylinder in exactly the same manner, and to the same combination.

Rekeying 780 Deadbolt
with a control key

When making a new key, also make a new control key with the same bitting (combination) as the new key.

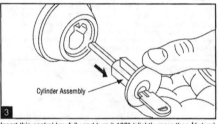

1
Deadbolt on door.

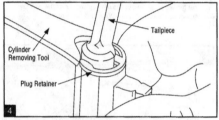

2
Locate the special "control key" which is cut specifically for the combination of your lock, but with a notch on the bottom of the key blade.

NOTE: If not available, cut a blank "control key" to match existing key.

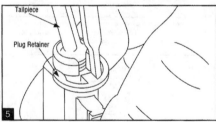

3
Insert this control key fully and turn it 120° (slightly more than 1/4 turn) counterclockwise. Pull the cylinder assembly straight out. Turn the key back to the vertical position and remove it.

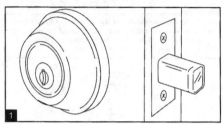

4
Lay the cylinder assembly face down on a flat surface. The tailpiece must be removed. To do this, fit the forked end of the Cylinder Removing Tool into the open end of the plug retainer and push it out. If the cylinder has a sleeve on the tailpiece, remove first and set aside.

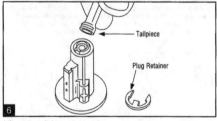

5
Remove the retainer completely using the other end of the tool.
Use care not to deform retainer.

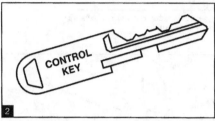

6
Lift out the tailpiece.

7

Insert old key or Control Key, turn 45° to the right or left. Keep key at this angle for next step.

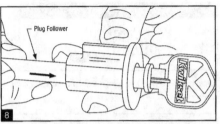

Plug Follower

8

Using the follower, push the plug out from the back. Make sure it stays tight against the plug or the top pins and springs will fall out (see page 19 if top pins and springs fall out). Also, make sure the follower sticks out at least one inch. **Leave it in place until you reassemble.**

Plug

9

Drop out the old pins, remove the old key and insert the new key. (We will assume you have already gauged the new key.)

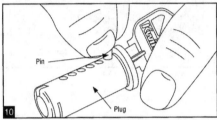

Pin

Plug

10

Begin inserting the new bottom pins from the rekeying kit according to the cut combination of the new key. You may handle the bottom pins with tweezers or your fingertips. As in our example (324625), use a #3 bottom pin for position one (which is a "3" depth on the key). A #2 bottom pin for position two, #4 pin for position 3, etc. Pins actually used will correspond with the key available.

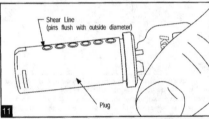

Shear Line
(pins flush with outside diameter)

Plug

11

If all the bottom pins are the proper size for the key, they will create the shear line. **Check this carefully.** A pin too long will prevent insertion of the plug while the key is inserted. A pin too short will allow the cylinder to lock up with no key operation, which will require complete disassembly.

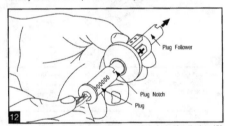

Plug Follower

Plug Notch

Plug

12

Begin reassembly in the same manner you started; position the key 45° to the right or left so that the top pins will not fall into the plug notch. Then use the plug to push the follower back through the cylinder.

13

Holding the plug firmly in the cylinder with your thumb, turn the key straight up and remove it.
**CAUTION: Do not allow the plug to come out as you
          remove the key.**

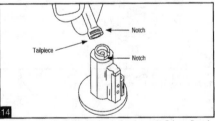

Notch

Tailpiece

Notch

14

Install the tailpiece with the notches in the plug and tailpiece aligned.

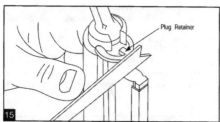

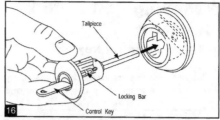

Use the cylinder removing tool to reseat the plug retainer. Then insert the key and test to make sure everything turns and operates properly. **If a tailpiece sleeve was used with your particular model and removed for rekeying, replace it now.**

Turn the new control key (cut to match new deadbolt key) about 120° counterclockwise until locking bar is flush with the cylinder body. Look into the cylinder hole to observe the orientation required for the tailpiece. Rotate the tailpiece accordingly.

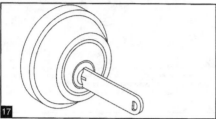

Insert cylinder assembly into deadbolt.

Model 780 (single cylinder) — Make sure the tailpiece aligns properly with latch hole and thumbturn hole.

Model 785 (double cylinder) — When inserting outside cylinder, turn control key to vertical position with cuts of key up. This will turn the solid tailpiece so it will not droop down. Insert cylinder assembly into lockset until it stops. Rotate the control key clockwise until locking bar is flush with cylinder. Then push cylinder in fully. Rotate tool clockwise to vertical position and withdraw. Check with new deadbolt key for proper function.

Rekeying 780 & 980 Deadbolt
without a control key

(If no control key exists, use the customer's existing key to cut one. When making the new key, also make a new control key with the same bitting (combination) as the new key.)

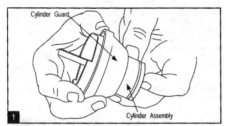

1 Remove cylinder assembly from cylinder guard.

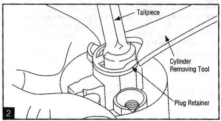

2 Disengage the plug retainer by pushing the forked end of the cylinder removal tool into the open end of the retainer and pushing it out. If the cylinder has a sleeve on the tailpiece, remove first and set aside.

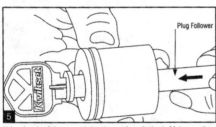

3 Remove the retainer completely using the other end of the Cylinder Removing Tool. **Use care not to deform retainer.**

Lift out the tailpiece.

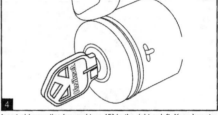

4 Insert old operating key and turn 45° to the right or left. Keep key at this angle for next step.

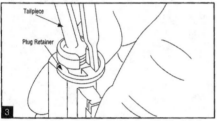

5 Using the plug follower, push the plug out from the back. Make sure the Plug Following Tool stays tight against the plug or the top pins and springs will fall out (see page 00 if this happens). Make sure the follower sticks out at least one inch. **Leave it in place until you reassemble.**

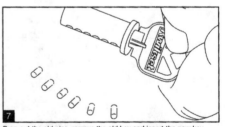

7 Drop out the old pins, remove the old key and insert the new key. (We will assume you have already gauged the new key.)

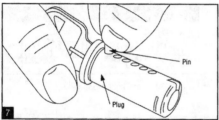

Begin inserting the new bottom pins from the rekeying kit according to the cut combination of the new key. You may handle the bottom pins with tweezers or your fingertips. As in our example (324625), use a #3 bottom pin for position one (a 3 depth on the key). A #2 bottom pin for position two, #4 pin for position three, etc. Pins actually used will correspond with the key available.

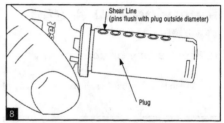

If all the bottom pins are the proper size for the key, they will create the shear line. **Check this carefully.** A pin too long will prevent insertion of the plug while the key is inserted. A pin too short will allow the cylinder to lock up with no key operation, which will require complete disassembly.

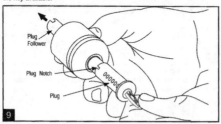

Begin reassembly in the same manner you started; position the key 45° to the right or left so that top pins will not fall into the plug notch, then use the plug to push the follower back through the cylinder.

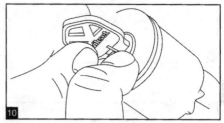

Holding the plug firmly in the cylinder with your thumb, turn the key straight up and remove it.
CAUTION: Do not allow the plug to come out as you remove the key!

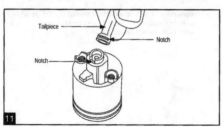

Install the tailpiece with the notches in the plug and the tailpiece aligned.

Use the cylinder removing tool to reseat the plug retainer. Then insert the key and test to make sure everything turns and operates properly. If a tailpiece sleeve was used with your particular model and removed for rekeying, replace it now.

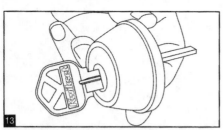

Insert the rekeyed cylinder back into the collar assembly. Rekeying is complete. If your customer has purchased a double cylinder model, simply rekey the other cylinder in exactly the same manner, and the same combination. Check for proper operation.

Rekeying 780 & 980 Deadbolt
without a key or control key

Step-by-step instructions for:

Procedure to remove cylinder assembly from housing. This also applies to TITAN NightSight products.

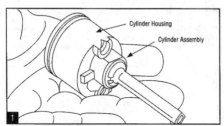

Complete cylinder assembly.

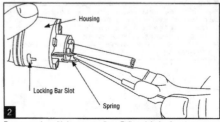

Remove spring with long nose pliers. Pull straight back, using care not to deform spring. Use the U-shaped spring you just pulled out to pick out the locking bar from the slot in the housing.

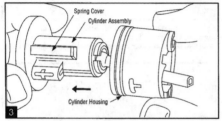

Remove cylinder assembly from housing. Then disassemble by removing spring cover and dropping out all pins and springs. Remove plug retainer and plug. Begin rekeying with new keys and reassemble locking bar and spring with housing, while using methods described in Section III step 12.

Notes: If a control key is made, the cylinder assembly can be installed into the housing easily. Follow instructions on rekeying with a control key.

However, if only a standard operating key is used, follow Section III instructions (rekeying if top pins and springs accidentally drop out.)

Rekeying 980 Deadbolts
What to do if the top pins and springs drop out during rekeying

This section deals with the situation when an accident occurs using a follower and one or more top pins and springs come out of the cylinder shell. Now, all the top pins, springs and spring cover must be removed.

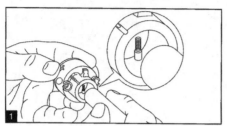

Pins and springs accidentally drop.

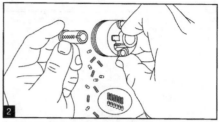

Pull plug completely out of cylinder shell allowing all top pins and springs to fall out. Carefully remove the spring cover. Do not deform it during removal or you will have to replace it with a new one.

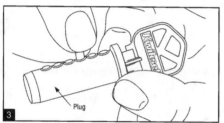

Insert the new key into the plug and add correct bottom pins to be flush with the shear line. (Outside surface of plug.)

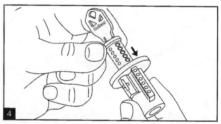

Insert the loaded plug (with key still inserted) into cylinder shell.

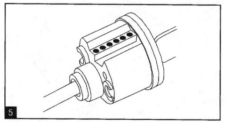

Insert/install the tailpiece and plug retainer.

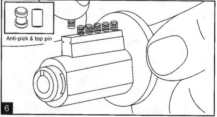

Add one top pin then one spring to each pin chamber. The anti-pick top pins should be placed in the first three pin chambers (1,2,3). By placing the conventional top pins (with full diameter) in the last chambers (4, 5, 6), this will provide smoother operation during the key insertion process. Conventional top pins will reduce plug rotation and maintain better chamber hole alignment between plug and cylinder body. Because tumblers 4, 5, and 6 are the last to be raised during key insertion, they will hold the plug during this process.

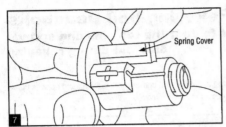

Install the spring cover by applying even pressure over springs until all four detent clamps are engaged.

Rekeying Knobs & Levers

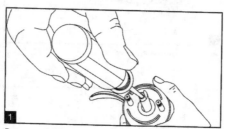

Remove product from box, look into end of spindle.

With a slotted screwdriver, turn spindle so that it is 45° to the half-round. Or. . .

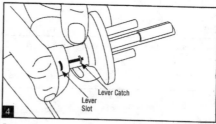

. . .if product is installed, turn button to 45°.

Depress lever catch (with a slotted screwdriver) to remove from chassis, if the lever catch does not depress easily, check your 45° angle.

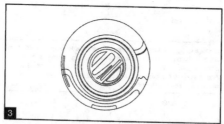

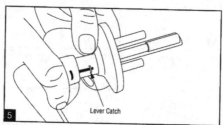

Slide knob or lever off the chassis. There are no screws to remove on the lever so just slide off interior shield and the cylinder will then come out.

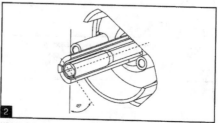

For Windham and Bedford knobs, slide the exterior sleeve off.

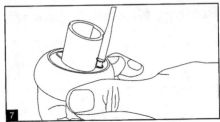

7

Using a slotted or torque screwdriver, remove both screws from the knob.

8

At this point, remove the interior sleeve.

9

Remove the cylinder from the knob. You can now re-key the lock.

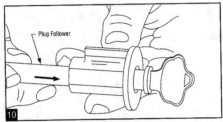

10

Remove C-clip and turn key off center. Using the follower, push the plug out from the back. Make sure it stays tight against the plug or the top pins and springs will fall out (see page 23 if top pins and springs fall out). Also, make sure the follower sticks out at least one inch. **Leave it in place until you reassemble.**

Plug Follower

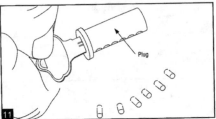

11

Drop out the old pins, remove the old key and insert the new key. (We will assume you have already gauged the new key.) Re-pin plug and reassemble the C-clip.

Plug

12

Place interior sleeve on top of the cylinder. Insert both screws firmly back into place.

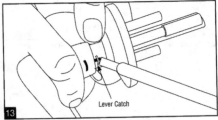

13

Place the exterior sleeve back into place.

Lever Catch

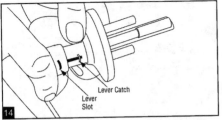

14

With spindle still in a 45 degree position, place back onto chassis.

Lever Catch
Lever
Slot

Step-by-step instructions for:

SOCIETY BRASS
C O L L E C T I O N

Rekeying Low Profile Deadbolt
with existing key

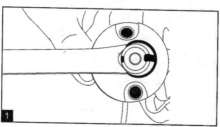

1 Remove the E-clip from the plug with the C-shaped side of your removal tool.

2 Turn key off vertical, so you do not allow the springs to release.

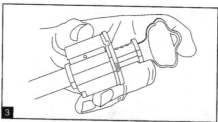

3 Use plug follower to remove plug.
(Note: The key pictured is the FINAL SBC key)

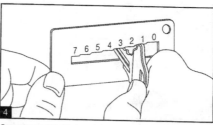

4 Gauge key and re-pin plug.

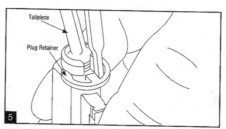

Tailpiece

Plug Retainer

5 After re-pinning, use plug follower to slide back into cylinder then replace E-clip with tailpiece in place (if the E-clip has spread apart too far, replace with new clip from kit).

Rekeying Low Profile Deadbolt

**without existing key
or if springs have fallen out**

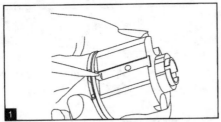

Slide spring cap back with awl or small slotted screwdriver.

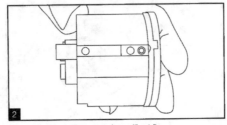

If you slide this slowly, your springs will not fly.

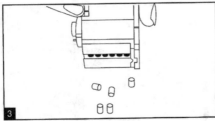

Remove springs and top pins, then remove E-clip and re-key plug.
After replacing pins and springs, slowly slide cap back on.

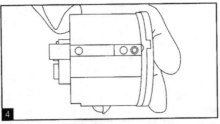

Load top pins (.180) and springs from top, slide spring cover back
on, pressing the springs down, one at a time, with a small slotted
screwdriver, as you slide cap over them.

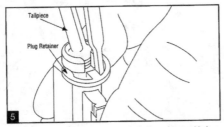

After re-pinning, replace plug and apply E-clip and torque blade.

7

Nonlocking
Door Hardware

Locks represent a small part of the hardware commonly used on or with doors. Other hardware includes hinges, strikes, and door reinforcers. A locksmith who is familiar with the wide range of door hardware can profit greatly by selling and installing them. This chapter reviews a wide range of door hardware.

Butts and Hinges

A door that binds or sags can be an eyesore and can hinder the performance of locks and alarms. Often these problems can be resolved by simply replacing the door's butts or hinges.

Many people use the terms "butt" and "hinge" synonymously. However, a distinction exists between the two. All butts are hinges, but all hinges aren't butts. Hinges designed for applications in which their leaves normally abut each other—such as on the edge of a door and door jamb—are called butt hinges, or butts for short. Usually one or both leaves of a butt are swaged, or bent slightly at the knuckles. That brings the leaves into closer contour with the barrel and allows for a tighter fit.

Butts and hinges come in a variety of shapes, styles, and types to be used for many different functions and applications. A butt or hinge usually consists of three basic parts: two metal leaves or plates—each with knuckles on one edge—and one pin that fits through the knuckles of both leaves. When the knuck-

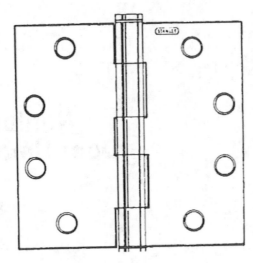

Figure 7.1   A butt hinge.

les of both leaves are joined, they form the barrel of the butt or hinge. See Figure 7.1.

The pin may be removable or nonremovable. A *removable pin* can be pulled out of the barrel. A *nonremovable pin* is typically retained by a small retaining pin or set screw. Ordinarily, the pin or set screw is concealed when the door is closed.

Nonremovable pins are especially useful for exterior doors where the butt or hinge is exposed to the outside. Without such pins, a person could enter a door simply by pulling the pins out of the door's hinges.

Fast, or rivet, pins have both ends machined on and are factory sealed. This type of nonremovable pin is often used in prisons and psychiatric hospitals.

Classification

Butts and hinges are classified in three ways: by screw hole pattern, by type of installation, and by function. Those that have standard screw-hole sizes and patterns are called *template hinges*. They're used mostly on metal doors and pressed metal frames.

Nontemplate hinges are those with staggered screw-hole patterns; such hinges are used on wooden doors and wood frames. *Blank face* or *plain hinges* have no predrilled screw holes. These are used when it's necessary to field-drill holes or weld hinges into place.

In addition to screw hole patterns, butts are classified by type of installation. *Full-mortise butts* are installed by mortising both leaves. See Figure 7.2. *Full-surface butts* are installed by surface mounting both leaves. Butts that have one leaf mounted to the door frame and the other mortised into the edge of the door are *half-mortise butts. Half-surface butts* are installed in the reverse manner. See Figure 7.3.

Sometimes butts and hinges are identified by how they function. For example, a "clear swing hinge" allows a door to swing

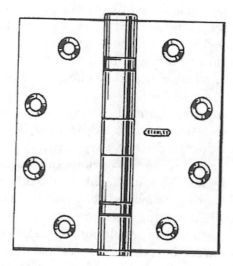

Figure 7.2   A full-mortise template hinge.

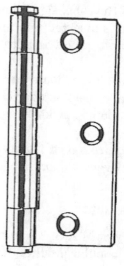

Figure 7.3   A half-surface is installed by mortising the jamb leaf.

clear of the passageway (usually a 180-degree swing), permitting full use of the door opening.

Regardless of the door type, weight, or size, a butt can usually be used on it. Butts come in four standard sizes, based on barrel length: 3-inch, 3-1/2 inch, 4-inch, and 5-inch. These come in a variety of finishes to match other door hardware. (To see a list of finishes, see Appendix A.) The following list describes many different kinds of butts and hinges.

Simple butts are nonhanded; they can be used on both left-hand and right-hand doors. These butts have nonremovable pins.

Loose-pin butts also have nonremovable pins. Their pins are fixed into one leaf. *Loose-pin butts* are designed to allow the door to be removed without disturbing the pin or unscrewing the leaves. The door is removed by lifting it up, so the leaves attached to the pins clear the fixed pins. Loose-pin butts are handed.

Rising butts are also handed. *Rising butts* have knurled knuckles and are designed to lift a door up as it swings open. They can allow a door to clear a heavy carpet.

Ball-bearing butts turn on two or more lubricated ball bearings, instead of pins. The lubricant and ball bearings are housed in "ball-bearing raceways" that look like small knuckles of the barrel. Ball-bearing butts are used on heavy doors.

Concealed hinges are used on folding doors and look different from most other types of hinges. They're installed by drilling two properly sized parallel holes, one in each meeting edge of the door. One cylindrical end of the concealed hinge is inserted into one hole, and the other end is inserted into the other hole. Then, both ends are screwed into place, resulting in a completely concealed hinge.

Gravity-pivot hinges permit folding doors to swing either way. *Spring-loaded hinges* can be used as door closers on fire doors and large screen doors. Some models have adjustable spring tension.

Pivot-framed hinges don't require a door frame. Such hinges are used for recessed, flush, or overlay doors.

M.A.G. Manufacturing History and Products

This section reviews the history of M.A.G. Manufacturing, and the company's newest and most popular products for locksmiths.

One evening in the late 1960s, Howard Allenbaugh, founder of M.A.G. Manufacturing, was listening to a custodian gripe about a recent rash of student break-ins at a local college. As the custodian complained about not knowing how to prevent students from kicking in locked doors. Howard sketched out a drawing of the now-famous *Door Reinforcer*, a U-shaped metal plate that is placed around the door knob and lock to strengthen the door at its weakest point, thus preventing kick-ins. See Figure 7.4. More than 30 years since its inception, the Door Reinforcer is now a popular security feature in hotels, restaurants, apartments, offices, and homes.

Since the design of the Door Reinforcer, the company grew from one to sixty employees in the U.S., and now offers over 1,000 security products. Located in southern California, M.A.G. is dedicated to providing consumers with innovative security and safety products, as well as fulfilling social and ethical responsibilities. The company is involved in many humanitarian efforts, including the national Safe At Home campaign, which provides financial support to assist child-

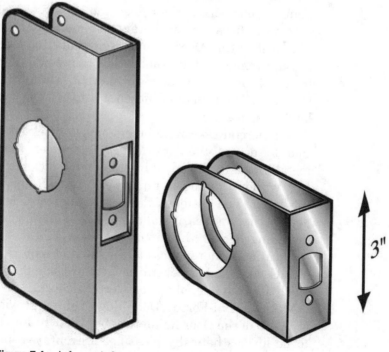

Figure 7.4 A door reinforcer.

safety advocacy groups in their efforts to educate parents on at-home safety practices.

Since the passing of Howard Allenbaugh in 2003, M.A.G. Manufacturing has been lead by the founder's son, Mark H. Allenbaugh—a practicing attorney, and a former business and professional ethics professor at George Washington University.

M.A.G.'s business was built by locksmiths seeking solutions to their workplace challenges. This same core relationship is the foundation of M.A.G.'s innovation and design strategy to be the locksmith's security and hardware provider. Providing commercial-grade security hardware remains at the forefront of M.A.G.'s focus, as it has been from since the company's inception.

M.A.G. Manufacturing regularly exhibits at several security conventions, including the Associated Locksmiths of America, the Door and Hardware Institute, and the Security Hardware Manufacturers Association.

The company's most popular security products among locksmiths are Door Reinforcers, Door Edge Guards, latch guards, Key Lock-Out, strikes, and filler plates. M.A.G. Manufacturing's Door Reinforcer comes in numerous finishes: polished brass, antique brass, satin bronze, oil-rubbed bronze, satin nickel, aluminum, polished stainless steel, and stainless steel.

M.A.G. offers residential and commercial Door Reinforcers. The residential Door Reinforcers come in several finishes: brass, chrome, and duranodic. The commercial models come in brass, chrome, duranodic, prime-coat, aluminum, and stainless steel. The primary difference between the two types is the latch preparation.

Commercial Door Reinforcers have a recessed edge preparation to accept the lock latch. This gives the professional installer the flexibility of interchanging locks or repairing the lock without having to remove the Door Reinforcer to access the latch. The residential Door Reinforcers have a flat-face design that makes installation of the plate much easier by installing it over the latch. However, this isn't intended for applications where access to the latch is required or probable. M.A.G. also makes several models of Door Reinforcers exclusively for commercial use on electronic locks.

M.A.G.'s Uni-Force® Door Edge Guard provides the same strength of the Door Reinforcer, but the Uni-Force® is designed to be an aesthetically pleasing, preventative product versus the Door Reinforcer, which in addition to adding an aesthetic element, also covers damage caused by kick-ins or covers the preparation marks for previous locks.

M.A.G.'s latch guards prevent someone from spreading the door and frame. There are models for both in-opening doors and out-opening doors. Residential latch guards come in three finishes: brass, chrome, and duranodic. Commercial models come in brass, chrome, duranodic, prime-coat, and aluminum.

Another popular M.A.G. security product is the *Key Lock-Out*, which is a temporary lock-out solution that uses the lockset keyway as the deterrent. Picture a key that has been broken off in the keyway obstructing any other keys from entering. But, with the Key Lock-Out, a patented tool allows this "broken" key bow to be removed when the need for the lock-out has passed or a professional installer changes the keyway. To engage the Key Lock-Out, simply insert the "key" into the lock and the detachable tip remains inside the keyway undetected. Once installed in a lock, it completely prevents a working key from being fully inserted into the lock. To remove the detachable tip, insert the "key" again and the tip reconnects and slides out, making the keyway operable.

"Key Lock-Out is an essential tool for many nerve-wracking situations, such as losing one's house keys," says Mark Allenbaugh. "Key Lock-Out is durable and can be used until lost keys are found, or until a locksmith can change the lock." Key Lock-Out is also an effective tool for apartment and building managers, construction sites, and business owners, as well as nursing homes and hospitals.

M.A.G. also offers several popular strikes. The company's *Adjust-a-Strike* is a T-Strike that uses tongue and groove plates to give the product its patented adjustable feature. This product is in great demand because of doors, frames, and entire buildings settling, which places the lock and strike out of alignment. The *T-Strike* lets you extend the strike past decorative moldings or masonry structures, allowing the latch to pass by without damaging the molding or masonry.

M.A.G.'s *Double-Strike* is designed to equally distribute strength between the deadbolt and knob latches.

M.A.G's newest security products are application-specific welded stud guards. The studs are concealed on the outside, making the latch guard more aesthetically pleasing. They come in various models specific to electronic locks, levers, mortise locks, electronic strikes, and key-in-knobs.

In 2006, M.A.G. introduced its M.A.G. PRO line of commercial-grade security hardware products, developed specifically for locksmiths.

M.A.G. doesn't sell to end users. Its products are only sold through locksmiths and other distributors, such as Ace Hardware, the Home Depot, True-Value, and Do It Best retail outlets.

M.A.G. Manufacturing offers Vendor Certification to locksmiths who demonstrate exceptional knowledge of the company's product lines. The certification process involves registering your company online, and then passing a short online quiz. To become certified, or to learn more about M.A.G. products, go to www.magmanufacturing.com.

8

Electricity and Electronics Basics

This chapter covers the minimum you need to know about electricity and electronics to install electronic security and safety systems. Part 1 focuses on electricity and Part 2 deals with electronics. If you've never installed a hardwired system (or if you have, but it didn't work right), read this chapter.

Electricity

There are three types of electricity: static, alternating current (ac), and direct current (dc). *Static electricity* happens in one place, instead of flowing through wires. An example is when you rub your shoes across a carpet, and then get a shock. You can also see static electricity work by rubbing a balloon on your hair, and then sticking the balloon to a wall. The most dangerous form of static electricity is lightning. For our purposes, the main concern about static electricity is preventing it from damaging electronic components.

Direct current comes from batteries. And, of course, *alternating current* comes from electrical outlets in a building. Both types work by following a continuous loop from the power source through conductors (usually wires) to a load (L)—such as an alarm or other electrical device—and back again. Current from batteries flows directly from the negative terminal back to the positive. AC, on the other hand, is more erratic; it "alternates" back and forth—first in one direction, and then in another along the circuit.

AC is generated at a power plant, and then transmitted many miles through a network of high-voltage power lines. Along the route, the electricity may be more than 750,000 volts. When it gets to the substation nearest you, a transformer is used to step-down the electricity to between 5,000 and 35,000 volts. Another step-down transformer on a nearby utility pole further reduces the electricity to about 240 volts, and that's carried to your building in a cable with two separate 120-volt lines.

Typically, buildings are wired so the two 120-volt lines work together at some outlets to provide 240 volts. The 120-volt circuits are for televisions, table lamps, and other small appliances. The 240-volt outlets are for large appliances, such as washing machines, clothes dryers, and refrigerators. (Some older homes have only one 120-volt incoming line.)

Most electronic security components are low voltage; they require much less than 120 volts. For those, you use the transformer that comes with them. You connect the component to the transformer, and then plug the transformer into a 120-volt outlet. That reduces the 120 volts going into the transformer to an amount that's right for the electronic device. The use of low-voltage transformers makes installing electronic security systems safe.

Types of Circuits

For electricity to do useful work, it needs to flow through a circuit. A *circuit* is the pathway, or route, of electric current. A series circuit has only one pathway; it has no branches. If multiple devices are wired in series, the current flows through each in turn, and a break at any device stops the flow for all the devices. An example is the old-style Christmas tree lights. When one goes out, they all go out. Figure 8.1 shows a *series circuit*. A *parallel circuit* has two or more pathways for electricity to move through. If multiple devices are wired in parallel, each is wired back to the power source, so each has its own current. See Figure 8.2. A *combination circuit* has both series and parallel portions.

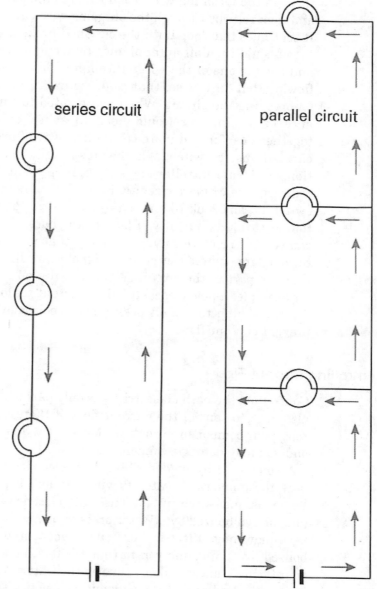

series circuit

parallel circuit

Figure 8.1   A series circuit.   **Figure 8.2**   A parallel circuit.

How a Circuit Works

The three basic parts of a circuit are a power source, conductors, and a load. The cord for an appliance has at least two wires. When you plug it into an electric outlet, electricity flows through the incoming wire (conductor) to the appliance (load), and then returns through the outgoing wire back to the outlet. It continues that loop until the circuit is broken.

You could turn off an appliance by using a pair of scissors and cutting one of the wires through which the electricity is flowing (but, of course, that could be dangerous). You would be breaking the circuit. When you wanted to turn the appliance on again, you would need to splice the wires back together. A safer and more convenient way of breaking a circuit is to use a switch. Whether it's a light switch, car ignition, or alarm-controller key switch, it opens and closes, or it redirects one or more circuits. From the outside, an installed switch doesn't look like much—just a little toggle, pushbutton, or turnkey. But, if you look at the back of a switch, you can see one or more tiny metal poles that move into position to complete or break circuits. The poles are little conductors that are part of the circuit. When you flip on a light switch, for example, a pole closes, to allow current to flow. When you flip the switch to the off position, the pole moves a little to prevent current flow.

Controlling Current Flow

Controlling the path of electricity is only part of the battle. You also need to control the current flow. If there's too little current, your components won't work correctly. Too much current and you may damage components.

A common way of describing how electricity is controlled uses the analogy of water flowing through a pipe (or hose). Water is measured in gallons and electricity in amperes (amps), symbolized by I. Water pressure is measured in pounds per square inch. Electrical pressure is measured in volts, symbolized by E. Imagine you have a 1,000-foot hose attached to your kitchen faucet and you turn on the water. When you finally stretch the hose all the way out, you notice the water isn't running out very fast. To make it run faster, you could plug the hose into a fire hydrant (not recommended) for more water pressure. Or, you could use a shorter hose.

Electric current can be controlled in similar ways. You can vary the pressure (voltage), and you can change the length and thickness of the conductors. Changing conductor size is one way of changing resistance, symbolized by R, to current flow.

Going back to the water hose, you could increase or decrease resistance by adding kinks to or removing them from the hose. (I don't know why anyone would do that. I'm just trying to make a point.) *Resistance*, measured in ohms and symbolized by Ω, is anything that slows the flow.

All conductors offer resistance to current flow, but some resist more than others. Copper, silver, and aluminum are good conductors, because they offer little resistance. Some materials, called insulators, are poor conductors. Examples of *insulators* include glass, dry wood, and plastic. Because plastic is a flexible insulator, it's used to sheath cable and wire to keep current from being misdirected … and to prevent you from getting shocked. Electrical tape, another insulator, is used to cover breaks in the plastic insulation. Plastic connectors are used to join and insulate bare ends of wire.

Here's another way of looking at current flow. Imagine you won a million dollars, and had to drive ten miles to pick it up. That money (or, rather, your desire for it) is the force pushing you to jump into your car, much like voltage pushes current. As you drive along, you are like current flowing. Various obstacles, such as bad weather, red lights, and police cars, are resistance. They slow you down. The fewer obstacles there are, the faster you can go (or flow). Suppose, instead of winning $1 million, you won only $10. That would be a less-motivating force, so you wouldn't drive as fast (or go as far) to get it. How far and how fast you'll go depends on your motivation and the obstacles (resistance) you face.

Ohm's Law

Having a general concept of how voltage, current, and resistance relate to one another is good, but you need to know how to put it into practical use. It would be a big hassle if every time you began an installation or were troubleshooting an electrical system, you had to resort to trial-and-error to determine what size power supply, as well as what material, length, and diameter of conductors, to use. Fortunately, you don't have to, because of a formula called Ohm's Law, created by the German physicist Georg Ohm.

Ohm's formula is remarkable and very simple. In a few minutes, you can have a good working knowledge of Ohm's Law.

In short, Ohm's Law says 1 volt can push 1 ampere through 1 ohm of resistance. Or, 2 volts can push 2 amperes through 1 ohm of resistance. This means if the resistance stays constant and you double or triple the voltage, the current also doubles or triples.

A common form of Ohm's Law is E = IR, or voltage in volts equals current in amperes times resistance in ohms. (IR means *I* times *R*. The multiplication sign (×) isn't used in these kinds of equations because it can be confused with the letter *X*, which has another meaning in mathematics.)

If you know the value of any two of the variables, you can find the third variable. Variations of Ohm's Law include I = E/R, or current equals voltage divided by resistance, and, R = E/I, or resistance equals voltage divided by current. Figure 8.3 shows a helpful picture form of these equations.

Try these examples:

- If you're using a 6-volt battery in a 3-ampere circuit, how much resistance is in the circuit? To find resistance, just divide the volts by current: E/I = 6/3, or R = 2 ohms. If a circuit has 10 ohms of resistance and 5 amperes of current, how much voltage is present? To find voltage, multiply current times resistance: E = 50 volts.

- If you have a 6-volt power source connected to a 4-ohm conductor, how much current is flowing through the conductor? To find current, divide voltage by resistance: I = E/R, so E = 1.5 amperes. (If you answered 0.666 ampere, you made the mistake of dividing resistance by voltage.)

- If you have an appliance with 15 ohms of resistance plugged into a 120-volt outlet, how much current will it draw? Because I = E/R, we know that I = 8 amperes. (If you

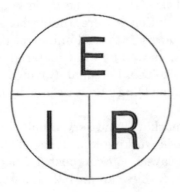

Figure 8.3 A picture version of Ohm's Law.

answered 0.125 ampere, you divided resistance by voltage. Try again.)

Basics of Electronics

Understanding common schematic drawings can help you read electrical drawings of architects, electrical designers, and security equipment manufacturers. You often find the drawings used in installation manuals.

Schematic Symbols

Hundreds of circuit components may go into an electronic device. To make it easier to show what components are in a device, each component is represented by a symbol, a letter, or a number. Although not every electronic technician uses the same symbols, their differences are so slight that, if you understand one, you won't likely get confused when you see another symbol for the same component.

Connecting wires (used as connectors in the circuit) are always shown as straight solid lines. They bend at sharp angles, usually 90 degrees. They rarely curve, except when crossing lines would make it unclear whether the wires are electrically connected. When no electrical connection exists between crossing lines, sometimes a half circle is drawn in the top line to show it jumping over the bottom line. That's the best way to do it. But some people do make straight lines cross each other when the wires aren't meant to be electrically connected. When you see one straight line crossing another, you need to find out what connection is being used.

Resistors

A *resistor* is one of the most common components used in electronics: it restricts the flow of electric current and produces a voltage drop. Its value is color-coded by four bands, based on the standard of the Electronics Industries Alliance (EIA). As Table 8.1 shows, the meaning of each band depends on its position and color. The band closest to the end of the resistor represents the first position number, depending on its color. The second band represents the second position number. The third band represents the multiplier. The fourth band is used to show the tolerance of the resistor. For example, say you have a band that's marked from the position closest to the end: red, violet, orange, and silver. Using Table 8.1, you can see the first number is 2 and the second number is 7, which are the first and second position numbers, respectively. That means the number is 27. The third position number, red, means the multiplier is 1,000. When you multiply 27 by 1,000, you get the nominal resistor value of 27,000 ohms. The resistor manufacturing process isn't perfect, and there's always some tolerance. The last color band tells how much above or below the nominal resistor value the actual value may be. The silver band in the last position means the actual resistor value is in the range of ±10 percent of the nominal value. That doesn't exactly mean the resistor is more or less than its nominal value—it could be exactly the same value.

There are several types of resistors. *Fixed resistors* have a fixed value in ohms. *Variable resistors* (also called rheostats or potentiometers) can be adjusted from zero to their full value to alter the amount of resistance in a circuit. *Tapped resistors* are a cross between a fixed and a variable resistor. Like a variable resistor, tapped resistors can be adjusted. Once adjusted, how-

TABLE 8.1 Color-coded bands based on the standard of the Electronics Industries Alliance (EIA)

Color	Band 1	Band 2	Band 3	Band 4
Black	0	0	1	—
Brown	1	1	10	1%
Red	2	2	100	2%
Orange	3	3	1,000	3%
Yellow	4	4	10,000	4%
Green	5	5	100,000	—
Blue	6	6	1,000,000	—
Violet	7	7	10,000,000	—
Gray	8	8	100,000,000	—
White	9	9	—	—
Gold	—	—	0.1	5%
Silver	—	—	0.01	10%
No color	—	—	—	20%

ever, tapped resistors become like a fixed resistor and cannot be adjusted anymore.

Resistance is also shown by the letter R, and sequential numbers if there are multiple resistors. For example, R1, R2, and R3 might appear at various points along a circuit diagram near schematic drawings of resistors, indicating they're the first, second, and third resistors listed on the schematic diagram. Sometimes a resistance value is also shown on a diagram, using the Greek letter omega (Ω). For instance, you might see 30 kΩ near a schematic symbol for a resistor.

Capacitors

Like resistors, capacitors are among the most common components used in electronics. The *capacitor* stores electricity and acts as a filter. It allows alternating current to flow, while blocking direct current. A capacitor consists of two metal plates or electrodes separated by some kind of insulation, called *dielectric*, such as air, glass, mica, polypropylene film, or titanium acid barium. The type of dielectric used affects how much capacitance can be obtained relative to the size of the capacitor, and in which applications it can best be used.

There are two types of capacitors: fixed and variable. The tuning dial of a radio is generally attached to a fixed capacitor. When you turn the dial, you're changing its frequency. There are many schematic symbols for identifying different types of capacitors. The letter C is used to refer to all types of capacitors, however. A whole number next to the letter C shows multiple capacitors exist and which one is within the circuit. The value, or capacitance, of the capacitor, shown in microfarads, may also be near the schematic symbol.

Switches

As mentioned earlier in this chapter, one way to turn a light on and off is to cut a wire that's part of the circuit, and then splice the wires back together when you want to turn the lamp on again. Using a switch is much safer and more convenient. Whenever you turn on a light or start your car, you're moving a switch into position to connect the circuit. A switch can have one or more contact points to allow multiple paths of current flow. Typically, switches are designed so you don't see the tiny poles that move to complete or break a circuit. If you were to take a switch apart, you would see them.

Switches are illustrated and named according to how many poles they have, and whether the poles move separately or together. A *single-pole single throw switch* is a basic on/off type switch. Its pole makes a connection with one of two contacts. On a schematic diagram, switches are identified by the letter S. In cases of multiple switches, the S is next to a number showing its listing in the circuit.

Relays

A *relay* is a special type of switch that's operated electrically or electronically, instead of manually. A relay is like several automatic switches rolled into one. In a schematic diagram, a relay is identified by the letter K, and its contacts are usually numbered. The contacts are also identified as normally open, shown as NO, or normally closed, shown as NC. A *normally open contact* is open when no electrical power is applied. A *normally closed contact* is closed when no power is applied. A *solenoid* is a type of relay that uses a magnetic field to move a plunger or an arm.

Appendix D shows common schematic drawings for various electronic components.

9

Emergency Exit Devices

To comply with building and fire codes, businesses and institutions often have to keep certain doors as emergency exits, which can be easily opened by anyone at any time. (This is to help prevent not having enough quick ways out during a fire or other emergency.) In some cases, however, those doors that must remain easy to exit also need to be secured from unauthorized use (such as when the door may allow shoplifters to slip out unnoticed).

Most institutions and commercial establishments use emergency door devices as a cost-effective way to handle both matters. Such devices are easy to install and offer excellent money-making opportunities for locksmiths.

Typically, such devices are installed horizontally about 3 feet from the floor. They have a bolt that extends into the door frame to keep the door closed. They also usually incorporate either a push-bar or a clapper arm that retracts the bolt when pushed.

Some models provide outside key and pull access when an outside cylinder and door pull are installed. In these models, entry remains restricted from both sides of the door until the deadbolt is relocked by key from inside or outside the door.

Many emergency exit door devices feature an alarm that sounds when a door is opened without a key. The better alarms are *dual piezo* (double sound).

Other features to consider on an emergency exit door device include on which hand it's installed (nonhanded models are the most versatile), the length of the bolt (a 1-inch throw is the minimum desirable length), and special security features (such as a hardened insert in the bolt).

The rest of this chapter covers the Alarm Lock Pilferguard, courtesy of Alarm Lock Corp.

Installation and Operating Instructions
Model PG-10
PILFERGUARD

Installation

1. **To Remove Cover:** Depress test button and lift cover out of slot.

2. Mark and drill holes as per template directions and drill sizes.
   (5 for alarm unit, 2 for magnetic actuator)

3. For outside cylinder installation where required (go to Step 4 if not).
   A. Drill a 1¼" hole as shown on template.
   B. Install a rim type cylinder through the door and allow flat tailpiece to extend 1" inside door.
   C. Position cylinder so that keyway is vertical (horizontal if PG-10 is installed horizontally).
   D. Hold PG-10 in position over mounting holes and note that outside cylinder tailpiece is centered in clearance hole in base of PG-10 (rotate cylinder 180° if not).
   E. Tighten outside cylinder mounting screws.

4. Install PG-10 and magnetic actuator with 7 screws.

5. Install threaded (mortise) cylinder (1¼" long) in PG-10 cover using hardware supplied (see Figure 1). Key way must be horizontal so that tailpiece extends towards center of unit when key is turned.

6. Move slide switch to "OFF" (see Figure 2).
   Connect battery.
   Hook cover on end slot and secure with two cover screws.
   Note: One of these screws acts as tamper alarm trigger so be sure screws are fully seated.
   This completes the installation, procede to "check-out".

 **ALARM LOCK**
345 Bayview Avenue, Amityville, New York 11701

825-409-002

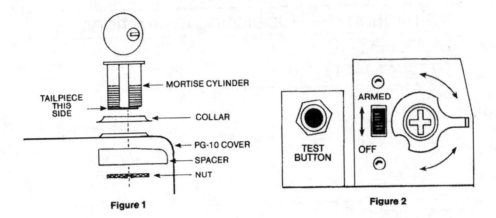

Figure 1   **Figure 2**

Check-Out and Operation

1. With slide switch in "OFF" position, depress test button – horns should sound.

2. To test using magnetic actuator:
   A. Close Door.
   B. Arm PG-10 by turning key clockwise 170 degrees.
   C. Open door, alarm should sound.
   D. Close door, alarm should remain sounding.
   E. Silence alarm by turning key counterclockwise until it stops.

3. Close door and re-arm PG-10 by turning key clockwise until it stops.

4. **Periodic Test:** Unit should be tested weekly using test button to insure battery is operational.

   Note: Test button only operates when PG-10 is turned off.

Special Conditions

Steel Frames - It is sometimes necessary on steel frames to install a non-magnetic shim between the magnetic actuator and the frame. This is done to prevent the steel frame from absorbing the magnets' magnetic field, which could cause a constant alarm condition or occasional false alarms.

The shim should be ½" by 2½" by ⅛" thick and may be constructed from plastic, bakelite or aluminum.

D.R.

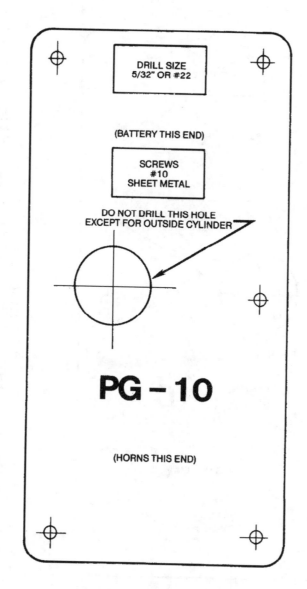

DRILL SIZE
5/32" OR #22

(BATTERY THIS END)

SCREWS
#10
SHEET METAL

DO NOT DRILL THIS HOLE
EXCEPT FOR OUTSIDE CYLINDER

PG – 10

(HORNS THIS END)

FOLD UP ON SOLID LINE AND
HOLD AGAINST DOOR AND STOP

MAGNETIC
ACUTATOR

MARK FOR ALARM
AND ACTUATOR

ONLY

INSTALLATION TEMPLATE

FOR SINGLE OUTSWINGING DOORS
RIGHT OR LEFT HAND – ALSO TOP MOUNT

ONLY

ALARM LOCK CORPORATION
333 BAYVIEW AVE. AMITYVILLE, N. Y. 11701

PS203

D.R.

LAY FLAT ON DOOR WITH SOLID LINE
EDGE OF INACTIVE DOOR (OR DOOR FRAME)

MAGNETIC
ACUTATOR

DO NOT FOLD UP
USE FLAT ON DOOR

DRILL SIZE
5/32" OR #22

(BATTERY THIS END)

SCREWS
#10
SHEET METAL

DO NOT DRILL THIS HOLE
EXCEPT FOR OUTSIDE CYLINDER

PG – 10

(HORNS THIS END)

ONLY

INSTALLATION TEMPLATE

FOR PAIRS OF OUTSWINGING DOORS
FOR PAIRS OF INSWINGING DOORS
FOR SINGLE INSWINGING DOORS
RIGHT OR LEFT HAND

ONLY

ALARM LOCK
345 Bayview Avenue
Amityville, New York 11701
For Sales and Repairs, (800) 645-9445
For Technical Service, (800) 645-9440
© ALARM LOCK 1994

PG21 & PG21E

INSTALLATION INSTRUCTIONS

WI738B 08/94

DESCRIPTION

The PG21 and PG21E are surface-mounted microprocessor-controlled door alarms. The PG21 is a basic battery-operated unit. The P021 E has provisions for an external power supply; external reed-switch contacts for multiple-door monitoring; and a built-in Form-C relay for connection to a fire or burglary alarm panel or other device.

The units mount on the interior of the doorwith a magnetic actuator on the frame; or, if external wiring is required (PG21 E only), vice versa. Full clockwise rotation of the installed standard mortise cylinder (not supplied) will alternately arm and disarm the unit. A selectable Annunciator feature beeps to signal opening of the door while the unit is disarmed. Opening the door, removal of the coveror any attempt to defeat the device with a second magnet, when armed, will activate the alarm. The units may be operated from outside the door with the addition of a standard rim cylinder. A test button and LED are used to check status; if the unit is armed, pressing the button will light the LED.

SPECIFICATIONS

Dimensions: 9 x **21/2** x 2W8" (22.8cm x 6.3cm x 6.8cm)(LxWxD)

Finish: Metallic Silver (PG21 MS, PG21 E); Metallic Bronze (P021MB); Red (PG21RD)

Power Requirements: 9-Volt Alkaline Battery (supplied). (The PG21E may be used with Model PP100 (optional) or other power supply providing 9 - 12Vdc at 500mA.)

Battery Life: Continuous Alarm, 3 hours; 2-Minute Shut down, see Table 1 below.

Alarms / Year	Battery Life (Months)
1	17.6
2	17.4
3	17.1
4	16.8
5	16.6
6	16.3
7	16.1
8	15.8
9	15.6
10	15.4

Table 1. Battery Life with 2-Minute Shutdown option.

Sounding Device: Piezo electronic sounder, sweep siren, 110dB at 10 feet
Shipping Weight: 1 lb, 10 oz.

INSTALLATION

Note: In many applications, the. need for a template may be eliminated by using the Magnet Alignment procedure in step 11. If outside key control or remote wiring is required, however, the template must be used.

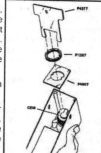

1. Referring to the illustration at right,

(a) install mortise cylinder (Alarm Lock Model CEM, optional) into the cover from the outside with the key slot at the 6 o'clock position;

(b) install the Spacer (P4607), then screw the Lock Ring (P1267) onto the Cylinder as shown and tighten using the SpannerWrench (P4577) supplied. Fig. 1. Installing mortise cylinder.

Fig. 1. Installing mortise cylinder.

2. If using a template, select the proper template for the specific type of door. (If not using a template, proceed to step 6.)

3. Mark and drill 9/16" holes per template directions on the door and jamb. (Four holes for baseplate and two holes for magnetic actuator.) Note: Certain narrow-stile doors require only two holes for mounting plate.

4. Model PG21E only: Mark and drill Whole per template for remote wiring to a control panel.

5. For outside key control only: Drill 1/4" hole as shown on template. Install a rim-type cylinder (not supplied) through the door and allow the flat tailpiece to extend 5/16" beyond the door. Position cylinder, keeping key slot pointing down (6 o'clock position) and tighten cylinder screws.

6. Knock out the necessary holes from the baseplate and install on the door with #8 sheet-metal screws supplied. If an outside cylinder is used, make sure that rim cylinder tailpiece fits into cross slot of ferrule.

7. Connect the battery; a chirp will sound. (This ensures that power is properly connected.) Important: Press the small CLEAR button at the lower-left corner of the circuit board (see Fig. 2).

8. Remove the jumper only on the side of the unit that the magnet will not be installed.

9. Select jumper options (J1- J4) as follows. Refer to Fig. 2. The unit is factory supplied with the Annunciator option enabled (J3 installed) and 2-Minute Shutdown alarm selected (J4 removed).

Fig. 2. CLEAR Button and Jumpers.

Important: Changes in the jumper configuration do not become effective until the unit is subsequently armed.

J1: Entry Delay. Alarm will sound 15 seconds after any unauthorized entry if unit is armed. This feature is used for authorized entry. Disarming within 15 seconds will prevent an alarm. To enable Entry Delay, install JI across both pins.

J2: Exit Delay. Unit will be activated after 15 seconds each time unit is armed to allow authorized exit without an alarm. To enable Exit Delay, install J2 across both pins.

J3: Annunciator. Unit will sound for 2 seconds whenever the door is opened while disarmed. To disable the Annunciator, remove J3.

J4: Continuous/2-Minute Shutdown. With J4 on, an alarm will sound continuously until the battery is depleted. With J4 off, the alarm will silence after two minutes and the unit will rearm (if the door is closed). The LED will start flashing to indicate that an alarm has occurred (Alarm Memoiy). Alarm Memory is cleared after about 4 hours or when the unit is rearmed. **Note**: If the door is still open after two minutes, the alarm will restart.

Note: Jumper J4 is connected at the factory to only one pin of the 2-pin connector (off). To select Continuous alarm, remove J4 and reinstall it across both pins.

10. Install the cover onto the baseplate, checking that the slide switch on the PC board fits into the cam hole. Secure the cover with #6-32 screws supplied. **Note**: The longest screw is the tamper screw. If not using a template, do not install the tamper screw until the magnet has been aligned and secured as follows.

11. (If the unit has been installed with the aid of a template, proceed to step 12.) After power-up, but before the tamper screw is installed, the unit will be in the Magnet Alignment Mode.

a. Place the magnet against the wall, adjacent to the bottom of the unit. Slowly slide the magnet upward. The LED will come on, indicating closure of the reed switch, then go out. Mark the door jamb at the bottom of the magnet. (**Note**: Sliding the magnet still further will cause the LED to light again; ignore subsequent indications.)

b. Similarly, place the magnet against the wall, adjacent to the top of the unit. Slowly slide the magnet downward. The LED will come on, then go out once again (see Note in step 11a). Mark the doorjamb at the top of the magnet.

12. Install the magnetic actuator so that it is centered between the two marks. Use the two #8 sheet-metal screws supplied. **Note**: On steel frames, it is sometimes necessary to install a non-magnetic shim between the magnetic actuator and the frame. This is done to prevent the steel frame from absorbing the magnet's magnetic field, which could cause a constant or occasional false alarm condition. The shim should be 1/2" x 21/2" x 1/8" thick non-magnetic material such as plastic, bakelite or rubber.

STANDBY MODE

After the magnet has been aligned and mounted, the tamper screw may be installed. This will place the unit in its regular Standby Mode (disarmed).

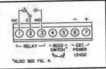

WIRING (PG1E only)

External wiring to the PG21E requires that the unit be mounted on the frame with the magnet on the adjacent door. (The PG21 E may be mounted on a door with the addition of a Model 271 Flexible Cable.) All wiring is made Fig. 3. Wiring Diagram. at its terminal strip. Wiring connections are summarized in the Wiring Diagram shown in Fig. 3.

Relay (Terminals 1-3). Terminal I = Normally Closed; Terminal 2 = Common; Terminal 3 = Normally Open.

Reed Switch (Terminals 4 & 5). Connect external reed-switch contacts to Terminals 4 and 5 and remove both jumpers. Note: If reed switches are wired in series as shown in Fig. 4, several doors may be monitored simultaneously. Using #22AWG wire, maximum wiring distance should not exceed 50 feet.

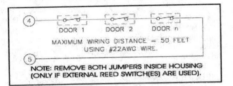

Power (Terminals 6 & 7). Connect an external power supply: positive (+) to Terminal 7; negative (—) to Terminal 6. Leave the internal battery in as a backup battery. If using the Alarm Lock PP100 Power Supply, connect the two battery clips to the unit and the internal battery as labeled.

FIELD TEST

1. With the door closed, turn the key fully clockwise, then release. A brief tone will sound indicating that the unit is armed (with or without delays, see options in INSTALLATION: step 9).

2. Push test button; the LED will come on for one second verifying that unit is armed. (If unit is disarmed, LED will not come on.) Note: If the test button is held down longer than 2 seconds, the sounder will be tested.

3. Open the door. A sweep siren alarm should sound.

4. Model PG21E only: On alarm, the relay will activate and remain latched for as long as the sounder is on. Relay contacts are rated at I 25Vac/dc at ½A.

5. To reset the alarm, turn key fully clockwise once again, then release. The unit is now disarmed.

Low Battery. In operation, whether armed or disarmed, the LED will blink about once every 4 hours when checking battery status. When the battery becomes weak, the unit will chirp and the LED will flash about once per minute to indicate the need for replacement.

EXPLODED VIEW

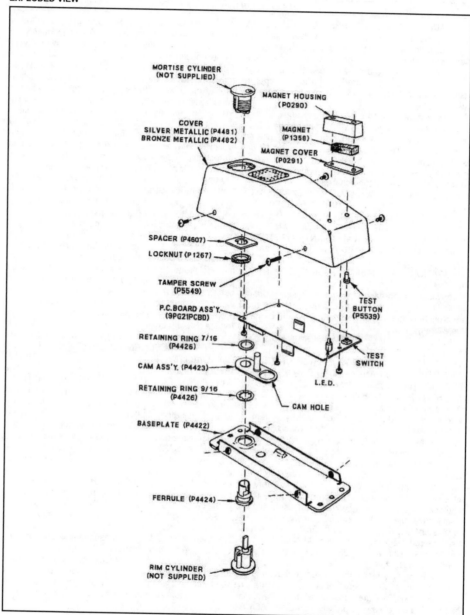

Fig. 5 Exploded view with part numbers

ALARM LOCK LIMITED WARRANTY

ALARM LOCK SECURITY SYSTEMS, INC. (ALARM LOCK) warrants its products to be free from manufacturing defects in materials and workmanship for 24 months following the date of manufacture. ALARM LOCK will, within said period, at its option, repair or replace any product failing to operate correctly without charge to the original purchaser or user.

This warranty shall not apply to any equipment, or any part thereof, which has been repaired by others, improperly installed, improperly used, abused, altered, damaged, subjected to acts of God, or on which any serial numbers have been altered, defaced or removed. Seller will not be responsible for any dismantling or reinstallation charges.

THERE ARE NO WARRANTIES, EXPRESS OR IMPLIED, WHICH EXTEND BEYOND THE DESCRIPTION ON THE FACE HEREOF. THERE IS NO EXPRESS OR IMPLIED WARRANTY OF MERCHANTABILITY OR A WARRANTY OF FITNESS FOR A PARTICULAR PURPOSE. ADDITIONALLY, THIS WARRANTY IS IN LIEU OF ALL OTHER OBLIGATIONS OR LIABILITIES ON THE PART OF ALARM LOCK.

Any action for breach of warranty, including but not limited to any implied warranty of merchantability, must be brought within the six months following the end of the warranty period.

IN NO CASE SHALL ALARM LOCK BE LIABLE TO ANYONE FOR ANY CONSEQUENTIAL OR INCIDENTAL DAMAGES FOR BREACH OF THIS OR ANY OTHER WARRANTY, EXPRESS OR IMPLIED, EVEN IF THE LOSS OR DAMAGE IS CAUSED BY THE SELLER'S OWN NEGLIGENCE OR FAULT.

In case of defect, contact the security professional who installed and maintains your security system. In order to exercise the warranty, the product must be returned by the security professional, shipping costs prepaid and insured to ALARM LOCK. After repair or replacement, ALARM LOCK assumes the cost of returning products under warranty. ALARM LOCK shall have no obligation under this warranty, or otherwise, if the product has been repaired by others, improperly installed, improperly used, abused, altered, damaged, subjected to accident, nuisance, flood, fire or acts of God, or on which any serial numbers have been altered, defaced or removed. ALARM LOCK will not be responsible for any dismantling, reassembly or reinstallation charges.

This warranty contains the entire warranty. It is the sole warranty and any prior agreements or representations, whether oral or written, are either merged herein or are expressly canceled. ALARM LOCK neither assumes, nor authorizes any other person purporting to act on its behalf to modify, to change, or to assume for it, any other warranty or liability concerning its products.

In no event shall ALARM LOCK be liable for an amount in excess of ALARM LOCK's original selling price of the product, for any loss or damage, whether direct, indirect, incidental, consequential, or otherwise arising out of any failure of the product. Seller's warranty, as hereinabove set forth, shall not be enlarged, diminished or affected by and no obligation or liability shall arise or grow out of Seller's rendering of technical advice or service in connection with Buyer's order of the goods furnished hereunder.

ALARM LOCK RECOMMENDS THAT THE ENTIRE SYSTEM BE COMPLETELY TESTED WEEKLY.

Warning: Despite frequent testing, and due to, but not limited to, any or all of the following; criminal tampering, electrical or communications disruption, it is possible for the system to fail to perform as expected. ALARM LOCK does not represent that the product/system may not be compromised or circumvented; or that the product or system will prevent any personal injury or property loss by burglary, robbery, fire or otherwise; nor that the product or system will in all cases provide adequate warning or protection. A properly installed and maintained alarm may only reduce risk of burglary, robbery, fire or otherwise but it is not insurance or a guarantee that these events will not occur. CONSEQUENTLY, SELLER SHALL HAVE NO LIABILITY FOR ANY PERSONAL INJURY, PROPERTY DAMAGE, OR OTHER LOSS BASED ON A CLAIM THE PRODUCT FAILED TO GIVE WARNING. Therefore, the installer should in turn advise the consumer to take any and all precautions for his or her safety including, but not limited to, fleeing the premises and calling police or fire department, in order to mitigate the possibilities of harm and/or damage.

ALARM LOCK is not an insurer of either the property or safety of the user's family or employees, and limits its liability for any loss or damage including incidental or consequential damages to ALARM LOCK's original selling price of the product regardless of the cause of such loss or damage.

Some states do not allow limitations on how long an implied warranty lasts or do not allow the exclusion or limitation of incidental or consequential damages, or differentiate in their treatment of limitations of liability for ordinary or gross negligence, so the above limitations or exclusions may not apply to you. This Warranty gives you specific legal rights and you may also have other rights which vary from state to state.

ALARM LOCK

INSTALLATION INSTRUCTIONS

PG30
Pilfergard Door Alarm

WI739B 12/97

DESCRIPTION

The PG30 is a keypad operated surface-mounted micro-processor-controlled door alarm. Three programmable security levels provide degrees of access to suit various applications, such as unattended fire stairways, airport security areas, delivery entrances, etc. The unit contains a Form-C relay as well as provisions for an external power supply and external reed-switch contacts.

The unit mounts on the door with a magnetic actuator on the frame; or, if external wiring is required, vice versa. Entering the Master Code will alternately arm and disarm the unit. A selectable Annunciator feature beeps to signal opening of the door while the unit is disarmed. Opening the door, removing the cover or attempting to defeat the unit with a second magnet, when armed, will activate the alarm. The unit may be operated from outside the door with the addition of a Model PG30KPD Remote Keypad (optional).

SPECIFICATIONS

Dimensions: 9" x 2½" x 2⅜" (22.8cm x 6.3cm x 6.8cm) (LxWxD)

Finish: Metallic Silver (PG30MS); Metallic Bronze (PG30MB)

Power Requirements: 9-Volt Alkaline Battery (supplied). (The PG30 may be used with Model PP100 (optional) or other power supply providing 9–12Vdc at 500mA.)

Battery Life: Continuous Alarm, 3 hours; 2-Minute Shutdown, see Table 1 below.

Alarms/Year	Battery Life (Months)
1	17.6
2	17.4
3	17.1
4	16.8
5	16.6
6	16.3
7	16.1
8	15.8
9	15.6
10	15.4

Table 1. Battery Life with 2-Minute Shutdown option.

Sounding Device: Piezo electronic sounder, sweep siren, 110dB at 10 feet

Shipping Weight: 1lb, 10oz.

FEATURES

- Three codes: Master, Management and Passage.
- Three levels of security to suit different applications.
- Three alarm types: sweep, steady or pulsing siren
- Two Annunciator volume levels.
- Automatic low-battery detection.
- Provisions for external power supply.
- Provsions for external reed-switch contacts for multiple-door monitoring.
- Built-in Form-C relay for connection to control panel or other device.

INSTALLATION

Note: In many applications, the need for a template may be eliminated by using the Magnet Alignment procedure in step 10. If outside keypad control or remote wiring is required, however, the template must be used.

1. If using a template, select the proper template for the type of door. (If not using a template, proceed to step 6.)

2. Mark and drill 9/64" holes per template directions on the door and jamb. (Four holes for baseplate and two holes for magnetic actuator.) **Note:** Certain narrow-stile doors require only two holes for mounting plate.

3. Mark and drill ⅜" hole per template for remote wiring to a control panel.

4. *For outside keypad control only:* Drill ¾" hole as shown on template. Refer to PG30KPD instructions for installation of outside keypad.

5. Knock out the necessary holes from the baseplate and install on the door with #8 sheet-metal screws supplied. If an outside keypad is used, make sure that it cable is routed through hole in baseplate.

6. Connect the battery; a chirp will sound. (This will confirm that power is connected properly.) *Important: Press the small* CLEAR *button at the lower-left corner of the circuit board (see Fig. 1).*

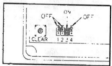

Fig. 1. CLEAR Button and Jumpers.

7. Remove the jumper only on the side of the unit that the magnet will *not* be installed.

8. Select jumper options (J1–J4) as follows. Refer to Fig. 1. The unit is factory supplied with the Annunciator option enabled (J3 installed) and *2-Minute Shutdown* alarm selected (J4 removed).

Important: Changes in the jumper configuration do not become effective until the unit is subsequently armed.

1

J1: Entry Delay. Alarm will sound 15 seconds after any unauthorized entry through door if unit is armed. This feature is used for authorized entry. To avoid an alarm upon entry, disarm unit within 15 seconds. To enable Entry Delay, install J1 across both pins.

J2: Exit Delay. Unit will be activated after 15 seconds each time unit is armed to allow authorized exit without an alarm. To enable Exit Delay, install J2 across both pins.

J3: Annunciator. Unit will sound for 2 seconds whenever the door is opened while disarmed. To disable the Annunciator, remove J3.

J4: Continuous/2-Minute Shutdown. With J4 *on*, an alarm will sound continuously until the battery is depleted. With J4 *off*, the alarm will silence after two minutes and the unit will rearm (if the door is closed). The LED will start flashing to indicate that an alarm has occurred (*Alarm Memory*). *Alarm Memory* is cleared after about 4 hours or when the unit is rearmed. **Note:** If the door is still open after two minutes, the alarm will restart.

Note: Jumper J4 is connected at the factory to only *one* pin of the 2-pin connector (*off*). To select *Continuous* alarm, remove J4 and reinstall it across *both* pins.

9. Install the cover onto the baseplate. Secure the cover with #6-32 screws supplied. **Note:** The longest screw is the tamper screw. *If not using a template, do not install the tamper screw until the magnet has been aligned and secured as follows.*

10. *(If unit has been installed using a template, proceed to step 11.)* After power-up, but *before* the tamper screw is installed, the unit will be in the *Magnet Alignment Mode*.

a. Place the magnet against the wall, adjacent to the *bottom* of the unit. Slowly slide the magnet *upward*. The LED will come on, indicating closure of the reed switch, then go out. Mark the door jamb at the *bottom* of the magnet. (**Note:** Sliding the magnet still further will cause the LED to light again; ignore subsequent indications.)

b. Similarly, place the magnet against the wall, adjacent to the *top* of the unit. Slowly slide the magnet *downward*. The LED will come on, then go out once again (see Note in step 10a). Mark the door jamb at the *top* of the magnet.

11. Install the magnetic actuator so that it is centered between the two marks on the door jamb. Use the two #8 sheet-metal screws supplied. **Note:** On steel frames, it is sometimes necessary to install a non-magnetic shim between the magnet and the frame. This is done to prevent the steel frame from absorbing the magnet's magnetic field, which could cause a constant or occasional false alarm condition. The shim should be ½" x 2½" x ⅛" thick non-magnetic material (plastic, bakelite, rubber, etc.).

STANDBY MODE

After the magnet has been aligned and mounted, the tamper screw may be installed. This will place the unit in its standby mode (disarmed).

ARMING & DISARMING

The unit is alternately armed and disarmed by entering the Master Code.

WIRING

External wiring to the PG30 requires that the unit be mounted on the frame with the magnet on the adjacent door. (The PG30 may be mounted on a door with the addition of a Model 271 Flexible Cable.) All wiring is made at its terminal strip. Wiring connections are summarized in the Wiring Diagram shown in Fig. 2.

2

Relay (Terminals 1–3). Terminal 1 = Normally Closed; Terminal 2 = Common; Terminal 3 = Normally Open.

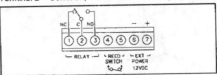

Fig. 2. Wiring Diagram. Also see Fig. 3.

Reed Switch (Terminals 4 & 5). Connect external reed-switch contacts to Terminals 4 and 5 and *remove both jumpers*. **Note:** If reed switches are wired in series as shown in Fig. 3, several doors may be monitored simultaneously. Using #22AWG wire, maximum wiring distance should not exceed 50 feet.

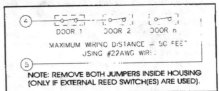

Fig. 3. Multiple door monitoring.

Power (Terminals 6 & 7). Connect an external power supply: positive (+) to Terminal 7; negative (−) to Terminal 6. Leave the internal battery in as a backup battery. If using the Alarm Lock PP100 Power Supply, connect the two battery clips to the unit and the internal battery as labeled.

FIELD TEST

1. With the door closed, enter a valid arming code. A brief tone will sound indicating that the unit is armed (with or without delays, see options in *INSTALLATION*: step 8).

2. Open the door. A sweep siren alarm should sound.

3. On alarm, the relay will activate and remain latched for as long as the sounder is on. Relay contacts are rated at 125Vac/dc at ½A.

4. To reset the alarm, enter the Master Code or the Management Code.

Low Battery. Whether armed or disarmed, the LED will blink about once every 4 hours when checking battery status. When the battery becomes weak, the unit will chirp and the LED will flash about once per minute, indicating the need for replacement.

PROGRAMMABLE FEATURES

The following features are software programmable.
- Codes: Master, Management, and Passage Codes;
- Security Level (Level 1, 2 or 3);
- Alarm Type: Sweep, Steady, or Pulsing Siren; and
- Annunciator Volume: Low or High.

It is recommended that the unit be programmed after it has been mounted and the door closed, as the tamper switch must be installed and the magnet properly positioned with respect to the active reed switch. If it is desired to preprogram the unit, keep the tamper switch closed (install tamper screw) and the reed switch closed (place magnet near switch) until programming is completed.

Note: (1) When entering codes or programming, each button must be pressed within 5 seconds of the last or the unit will time out. (2) Press the [AL] button to cancel a wrong code entry (prior to completion). (3) Seven beeps indicate that the programmed code has been accepted. (4) Five beeps indicate either a wrong code entry or a programming error. (5) Three unsuccessful attempts to enter a code will sound a loud 2-second alert.

Codes

Each code may be 3 to 5 digits in length.

Master Code. The Master Code can be used to program any code or feature or to arm or disarm. The default Master Code is 1-2-3-4-5. *This code must be changed before any other code or function can be programmed.* To program a new Master Code, press *[1] [2] [3] [4] [5] [AL] [1] [AL] [new Master Code] [AL]*.

Note: Be sure to enter the new Master Code carefully. If the wrong code is entered and is unknown, press the CLEAR button on the circuit board to reset the unit and clear all codes, then start over. (Whenever the CLEAR button is pressed, all programmed codes are erased and the Master Code defaults to 1-2-3-4-5.)

To program the Management Code (see below), enter the Master Code, then press *[AL] [2] [AL] [new Management Code] [AL]*.

Management Code. The Management Code may be used to program the Passage Code, any feature, or to silence the alarm.

Passage Code. This code allows passage without altering the arm/disarm status of the unit. If the unit is armed, the door must be closed within 15 seconds or the alarm will sound. The Passage Code may be used with any programmed security level.

To program the Passage Code, enter the Master Code or Management Code, then press *[AL] [3] [AL] [new Passage Code] [AL]*.

Security Level

There are three levels of security to accommodate different applications. **Note:** The tamper switch and keypad are disable until the door is closed. Security levels are programmable using the Master Code or Management Code:

Level 1. This provides the highest level of security and is the default mode of operation. Opening the door while armed will cause the alarm to latch on (after the entry delay, if enabled). The alarm may only be reset using the Master Code or the Management Code. To reprogram *Level-1* security (if changed), enter the Master Code or Management Code, then press *[AL] [3] [3] [AL]*.

Level 2. Entering the Passage Code will allow the door to be opened up to 15 seconds without activating an alarm. If no code is entered or if the door remains open longer than 15 seconds, a non-latching alarm will sound; that is, when the door is closed, the alarm will shut down and the unit will reset. Applications include unattended doors as may be found at fire stairwells, airport high-security areas, etc. that must remain closed after passage. To enable *Level-2* security, first enter the Master Code or Manage-

ment Code, then press *[AL] [4] [4] [AL]*.

Level 3. Entering the Passage Code allows the door to remain open indefinitely without activating an alarm. When the door is closed, the unit will rearm automatically. A typical application might be a delivery entrance where the door is propped open for long periods of time. To enable *Level-3* security, first enter the Master Code or Management Code, then press *[AL] [5] [5] [AL]*.

Alarm Type

Sweep Siren. This is the default alarm. To reprogram a Sweep Siren alarm (if changed), enter the Master Code or Management Code, then press *[AL] [1] [1] [AL]*.

Steady Siren. To program a Steady Siren alarm, enter the Master Code or Management Code, then press *[AL] [1] [2] [AL]*.

Pulsing Siren. To program a Pulsing Siren alarm, enter the Master Code or Management Code, then press *[AL] [1] [3] [AL]*.

Annunciator Volume

Low. This is the default setting. Low volume is adjustable by means of the potentiometer at the upper-left corner of the circuit board; clockwise rotation increases loudness. To reprogram Low-volume annunciator (if changed), enter the Master Code or Management Code, then press *[AL] [1] [4] [AL]*.

High. This setting provides maximum loudness. To program High volume, enter the Master Code or Management Code, then press *[AL] [1] [5] [AL]*.

EXPLODED VIEW

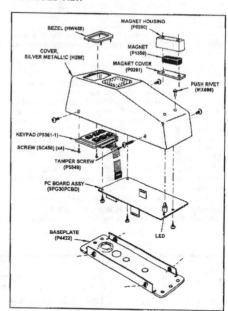

Fig. 4. Exploded view with part numbers.

ALARM LOCK LIMITED WARRANTY

ALARM LOCK SECURITY SYSTEMS, INC. warrants its products to be free from manufacturing defects in materials and workmanship for *twelve months* following the date of manufacture. ALARM LOCK will, within said period, at its option, repair or replace any product failing to operate correctly without charge to the original purchaser or user.

This warranty shall not apply to any equipment, or any part thereof, which has been repaired by others, improperly installed, improperly used, abused, altered, damaged, subjected to acts of God, or on which any serial numbers have been altered, defaced or removed. Seller will not be responsible for any dismantling or reinstallation charges.

THERE ARE NO WARRANTIES, EXPRESS OR IMPLIED, WHICH EXTEND BEYOND THE DESCRIPTION ON THE FACE HEREOF. THERE IS NO EXPRESS OR IMPLIED WARRANTY OF MERCHANTABILITY OR A WARRANTY OF FITNESS FOR A PARTICULAR PURPOSE. ADDITIONALLY, THIS WARRANTY IS IN LIEU OF ALL OTHER OBLIGATIONS OR LIABILITIES ON THE PART OF ALARM LOCK.

Any action for breach of warranty, including but not limited to any implied warranty of merchantability, must be brought within the six months following the end of the warranty period.

IN NO CASE SHALL ALARM LOCK BE LIABLE TO ANYONE FOR ANY CONSEQUENTIAL OR INCIDENTAL DAMAGES FOR BREACH OF THIS OR ANY OTHER WARRANTY, EXPRESS OR IMPLIED, EVEN IF THE LOSS OR DAMAGE IS CAUSED BY THE SELLER'S OWN NEGLIGENCE OR FAULT.

In case of defect, contact the security professional who installed and maintains your security system. In order to exercise the warranty, the product must be returned by the security professional, shipping costs prepaid and insured to ALARM LOCK. After repair or replacement, ALARM LOCK assumes the cost of returning products under warranty. ALARM LOCK shall have no obligation under this warranty, or otherwise, if the product has been repaired by others, improperly installed, improperly used, abused, altered, damaged, subjected to accident, nuisance, flood, fire or acts of God, or on which any serial numbers have been altered, defaced or removed. ALARM LOCK will not be responsible for any dismantling, reassembly or reinstallation charges.

This warranty contains the entire warranty. It is the sole warranty and any prior agreements or representations, whether oral or written, are either merged herein or expressly cancelled. ALARM LOCK neither assumes, nor authorizes any other person purporting to act on its behalf to modify, to change, or to assume for it, any other warranty or liability concerning its products.

In no event shall ALARM LOCK be liable for an amount in excess of ALARM LOCK's original selling price of the product, for any loss or damage, whether direct, indirect, incidental, consequential, or otherwise arising out of any failure of the product. Seller's warranty, as hereinabove set forth, shall not be enlarged, diminished or affected by and no obligation or liability shall arise or grow out of Seller's rendering of technical advice or service in connection with Buyer's order of the goods furnished hereunder.

ALARM LOCK RECOMMENDS THAT THE ENTIRE SYSTEM BE COMPLETELY TESTED WEEKLY.

Warning: Despite frequent testing, and due to, but not limited to, any or all of the following; criminal tampering, electrical or communications disruption, it is possible for the system to fail to perform as expected. ALARM LOCK does not represent that the product/system may not be compromised or circumvented; or that the product or system will prevent any personal injury or property loss by burglary, robbery, fire or otherwise; nor that the product or system will in all cases provide adequate warning or protection. A properly installed and maintained alarm may only reduce risk of burglary, robbery, fire or otherwise but it is not insurance or a guarantee that these events will not occur. CONSEQUENTLY, SELLER SHALL HAVE NO LIABILITY FOR ANY PERSONAL INJURY, PROPERTY DAMAGE, OR OTHER LOSS BASED ON A CLAIM THE PRODUCT FAILED TO GIVE WARNING. Therefore, the installer should in turn advise the consumer to take any and all precautions for his or her safety including, but not limited to, fleeing the premises and calling police or fire department, in order to mitigate the possibilities of harm and/or damage.

ALARM LOCK is not an insurer of either the property or safety of the user's family or employees, and limits its liability for any loss or damage including incidental or consequential damages to ALARM LOCK's original selling price of the product regardless of the cause of such loss or damage.

Some states do not allow limitations on how long an implied warranty lasts or do not allow the exclusion or limitation of incidental or consequential damages, or differentiate in their treatment of limitations of liability for ordinary or gross negligence, so the above limitations or exclusions may not apply to you. This Warranty gives you specific legal rights and you may also have other rights which vary from state to state.

Alarm Lock Systems
345 Bayview Avenue, Amityville, New York 11701
Call Toll Free: 800-ALA-LOCK (800-252-5625)

INSTALLATION TEMPLATE

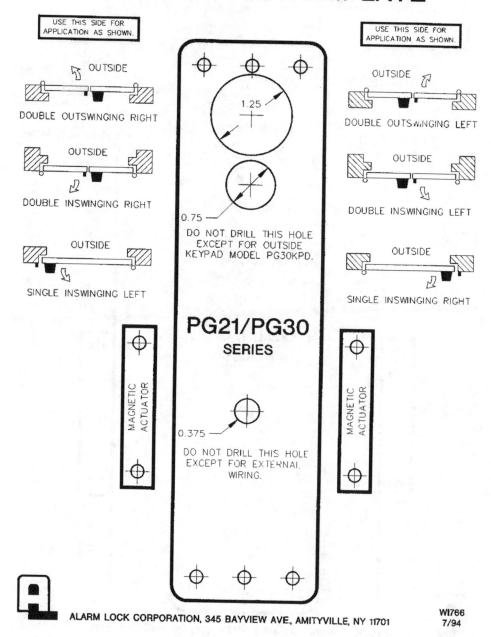

USE THIS SIDE FOR
APPLICATION AS SHOWN.

OUTSIDE

DOUBLE OUTSWINGING RIGHT

OUTSIDE

DOUBLE INSWINGING RIGHT

OUTSIDE

SINGLE INSWINGING LEFT

MAGNETIC ACTUATOR

1.25

0.75
DO NOT DRILL THIS HOLE
EXCEPT FOR OUTSIDE
KEYPAD MODEL PG30KPD.

PG21/PG30
SERIES

0.375
DO NOT DRILL THIS HOLE
EXCEPT FOR EXTERNAL
WIRING.

USE THIS SIDE FOR
APPLICATION AS SHOWN.

OUTSIDE

DOUBLE OUTSWINGING LEFT

OUTSIDE

DOUBLE INSWINGING LEFT

OUTSIDE

SINGLE INSWINGING RIGHT

MAGNETIC ACTUATOR

ALARM LOCK CORPORATION, 345 BAYVIEW AVE, AMITYVILLE, NY 11701

WI766
7/94

INSTALLATION TEMPLATE

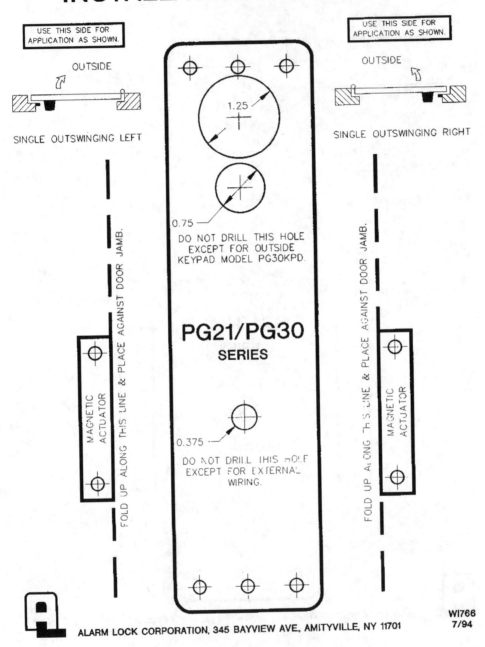

ADDENDUM
Model: PG30
Subject: Code Re-entry

Important: When attempting to silence the sounder after an alarm, if the sounder does not shut off due to improper or wrong code entry, wait 5 seconds before re-entering your code.

WI814

10/95

ALARM LOCK

INSTALLATION INSTRUCTIONS

PG30KPD
External Keypad

WI773A 8/95

The PG30KPD is an external keypad accessory to the PG30 Door Alarm that allows operation of the PG30 from either side of the door. It consists of a keypad with wall plate, a single-gang junction box, a ribbon-cable extension, and a 2-keypad adaptor module.

Note: (1) Use of the PG30KPD requires that the PG30 be mounted on the door frame with magnetic actuator on the door. (2) The PG30KPD is not intended for exterior use.

INSTALLATION

1. Remove the PG30 cover. Remove the 0.75" knockout in the baseplate just below the large (1.25") hole at the top. (Refer to the Installation Template WI766 furnished with the PG30.)

2. Remove an appropriate knockout in the single-gang box and install the box on the other side of the wall, close to the PG30 (within the limits of the extension cable).

3. Snake the ribbon extension cable through the wall so that the male end connects to the PG30KPD and the female end connects to either of the two male connectors on the 2-keypad adaptor module.

4. Inplug the PG30 keypad connector from the PG30 circuit board and plug it into the other male connector on the 2-keypad adaptor module.

5. Plug the 2-keypad adaptor module into the PG30 keypad connector on the circuit board.

6. Replace the PG30 cover. Install the PG30KPD wall plate onto the junction box using the two tamper-proof screws provided.

7. Check operation of both keypads.

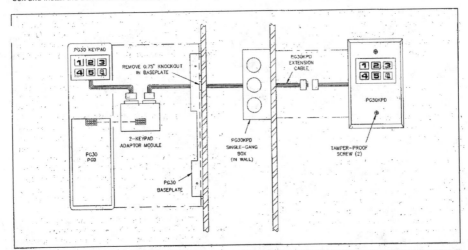

Alarm Lock Systems
345 Bayview Avenue, Amityville, New York 11701
Call Toll Free: 800-ALA-LOCK (800-252-5625)

ALARM LOCK LIMITED WARRANTY

ALARM LOCK Systems, Inc. (ALARM LOCK) warrants its products to be free from manufacturing defects in materials and workmanship for fifteen months following the date of manufacture. ALARM LOCK will, within said period, at its option, repair or replace any product failing to operate, without charge to the original purchaser or user.

In case of defect, contact the security professional who installed and maintains your security system. ALARM LOCK shall have no obligation under this warranty, or otherwise, if the product has been repaired by others, improperly installed, improperly used, abused, altered, damaged, subjected to accident, nuisance, flood, fire or acts of God, or on which any serial numbers have been altered, defaced or removed. ALARM LOCK will not be responsible for any dismantling, reassembly or reinstallation charges.

In order to exercise the warranty, the product must be returned by the user or purchaser, shipping costs prepaid and insured to ALARM LOCK. After repair or replacement, ALARM LOCK assumes the cost of returning products under warranty.

There are no warranties, express or implied, which extend beyond the description on the face thereof. There is no express or implied warranty of merchantability or a warranty of fitness for a particular purpose. Additionally, this warranty is in lieu of all other obligations or liabilities on the part of ALARM LOCK.

Any action for breach of warranty, including but not limited to any implied warranty or merchantability, must be brought within the six months following the end of the warranty period. In no case shall ALARM LOCK be liable to anyone for any consequential or incidental damages for breach of this or any other warranty, express or implied, even if the loss or damage is caused by the seller's own negligence or fault.

This warranty contains the entire warranty. It is the sole warranty and any prior agreements or representations, whether oral or written, are either merged herein or are expressly cancelled. ALARM LOCK neither assumes, nor authorizes any other person purporting to act on its behalf to modify, to change, or to assume for it, any other warranty or liability concerning its products.

In no event shall ALARM LOCK be liable for an amount in excess of ALARM LOCK's original selling price of the product, for the loss or damage, whether direct, indirect, incidental, consequential, or otherwise arising out of any failure of the product. Seller's warranty, as hereinabove set forth, shall not be enlarged, diminished or affected by, and no obligation or liability shall arise or grow out of, Seller's rendering of technical advice or service in connection with Buyer's order of the goods furnished hereunder.

ALARM LOCK RECOMMENDS THAT THE ENTIRE SYSTEM BE COMPLETELY TESTED WEEKLY.

Warning: Despite frequent testing, and due to, but not limited to, any or all of the following: criminal tampering electrical or communications disruption, it is possible for the system to fail to perform as expected. ALARM LOCK does not represent that the product/system may not be compromised or circumvented; or that the product or system will prevent any personal injury or property loss by burglary, robbery, fire or otherwise; nor that the product or system will in all cases provide adequate warning or protection. A properly installed and maintained alarm or lock may only reduce risk of burglary, robbery, fire or otherwise but it is not insurance or a guarantee that these events will not occur. CONSEQUENTLY, SELLER SHALL HAVE NO LIABILITY FOR ANY PERSONAL INJURY, PROPERTY DAMAGE, OR OTHER LOSS BASED ON A CLAIM THE PRODUCT FAILED TO GIVE WARNING. Therefore, the installer should in turn advise the consumer, and the consumer is hereby advised, to take any and all precautions for his or her safety including but not limited to, fleeing the premises and calling police or fire department, in order to mitigate the possibilities of harm and/or damage.

ALARM LOCK is not an insurer of either the property or safety of the user's family or employees, and limits its liability for any loss or damage including incidental or consequential damages to ALARM LOCK's original selling price of the product regardless of the cause of such loss or damage. If the user wishes to protect itself to a greater extent, ALARM LOCK will, at user's sole cost and expense, obtain an insurance policy to protect the user, supplemental to user's own policy, at a premium to be determined by ALARM LOCK's insurer upon written notice from user by Certified Mail, Return Receipt Requested, to ALARM LOCK's home office address, and upon payment of the annual premium cost by user.

This warranty shall be construed according to the laws of the State of New York. Some states do not allow limitations on how long an implied Warranty lasts or do not allow the exclusion or limitation of incidental or consequential damages, or differentiate in their treatment of limitations of liability for ordinary or gross negligence, so the above limitations or exclusions may not apply to you. This Warranty gives you specific legal rights and you may also have other rights which vary from state to state.

D.R.

10

Electric Strikes

An *electric strike* is architectural hardware designed to make frequently used doors more secure and convenient. These strikes are usually mounted in the frame of a door, and they use electricity either to hold or release a latch. A switch, which can be located close to or far away from the strike, can activate and deactivate an electric strike.

There are several popular manufacturers of electric strikes. This chapter reviews some models made by Adams Rite Manufacturing Company.

Strike Selection

Most Adams Rite strikes have flat faces. Two exceptions are the 7801 and the 7831 models, which have radius faces to match the nose shape of paired narrow-stile glass doors.

The basic Adams Rite size conforms to American National Standards Institute (ANSI) strike preparation, 1-1/4 by 4-7/8 inches. However, two other sizes are offered to fill or cover existing jambs or opposing stiles from previous installations of M.S. deadlock strikes (7830 and 7831) or discontinued series 002 electric strikes (7810).

Strikes are available with round corners for installation in narrow-stile aluminum, where preparation is usually done by router, and with square corners for punched hollow metal ANSI preparation or wood mortise.

Strikes with vertically mounted solenoids are designed to slip into hollow metal stile sections as shallow as 1.6 inches.

Horizontal solenoids for wood jambs require an easily bored 3-5/8 to 4-1/2-inch-deep mortise, depending on the model.

If the jamb was previously fitted with a 002 electric strike, the new 7810 unit will fit with minor alterations. If a hollow jamb was originally prepared to receive the bolt from an Adams Rite M.S. deadlock and you plan to substitute a 4710 latch, the 7830 (flat jamb) or 7831 (radius inactive door) will cover the old strike cutout.

The standard lip on all basic Adams Rite electric strikes accommodates 1-3/4-inch-thick doors that close flush with the jamb. If the door/jamb relationship is different, a long clip can be added if it is so specified.

Adams Rite electric strikes 7800 through 7831 mate with all Adams Rite series 4500 and 4700 latches. The 7840, 7840 ANSI, and 7870 are designed for mortise latches.

All standard operation strikes are unhanded and can be installed for either right- or left-hand doors.

The first electrical factor to determine is whether the operation is to be intermittent or continuous. If the door is normally locked and released only momentarily from time to time, it is *intermittent*. If the strike is rarely activated (unlocked) for long periods, the duty is *continuous*. A seldom-used requirement is *continuous/reverse action*, also called *fail safe*, in which the strike is locked only when its current is switched on.

For a normal intermittent application, specify an electric strike using 24 alternating current (ac) volts. This gives enough power for almost any entrance, even one with a wind-load situation. Yet this low-voltage range is below that requiring Underwriters Laboratories (UL) or Building Code supervision. Reliable transformers are available in this voltage.

The buzzing sound inherent in ac-activated strikes is usually not considered offensive in intermittent use. In fact, it acts as the "go" signal to a person waiting to enter. However, if silence is desired, specify a strike using 24 direct current (dc) volts. This dc current operates the strike silently. A buzzer or pilot light can be wired in if a signal is required.

Continuous duty is required when the strike will be energized for more than 60 seconds at a time.

Most continuous-duty applications can be supplied through the same 115/24-volt ac transformer used for intermittent jobs. The factory will automatically add those components necessary to achieve continuous performance at the voltage specified.

For long periods of unlocking, a Fail-Safe Reverse Action strike can be obtained. This might be required to provide the same service as a continuous-duty strike, but it will preserve current because it is on dc power. A Fail-Safe Reverse Action strike could also be used to provide a fail-safe unlocked door in case of power failure.

If a visual or other signal is required to tell the operator what the electric strike is doing, a monitoring strike is needed. This feature is specified by adding the proper dash number to the strike's catalog number. Two sensor/switches are added. One is activated by the latch-bolt's penetration of the strike and the other by the solenoid plunger that blocks the strike's release.

Low voltage for electric strike operation is obtained by using a transformer, which steps down the normal 115-volt ac power to 12, 16, 24, or other lower voltage. For this operation, three items must be specified: input voltage (usually 115 volts), output voltage (12, 16, or 24 volts), and capacity of the transformer, measured in volt-amps.

Skimping on the capacity of the transformer to save a few dollars will not provide adequate power for the door release. Adams Rite electric strikes for intermittent-duty models draw less than 1 amp (regardless of duty) and use a 4,602 current limiter, which stores electrical energy for high-use periods.

The wire must carry the electrical power from the transformer through the actuating switch (or switches) to the door release. It must be large enough to minimize frictional line losses and deliver most of the output from the transformer to the door release. For example, a small-diameter garden hose won't provide a full flow of water from the nozzle, particularly if it's a long run. Neither will an undersized wire carry the full current.

When you suspect insufficient electrical power in a weak door release, simply measure the voltage at the door release while the unit is activated. If the voltage is below that specified on the hardware schedule, the problem is in the circuit—probably an under-capacity transformer if the current length is short. A long run might indicate both a transformer and wire problem.

Frequently Asked Questions

Q: How do you define intermittent or continuous duty?
A: When there are only short periods of time during which the switch will be closed and the coil energized, the duty is classified as continuous.

Q: What is the electrical voltage source?

A: For the majority of electric strikes, 115-ac volt, 60-cycle alternating current is the power source. Unless there are specific customer demands to do otherwise, the recommendation is that the 115-volt, 60-cycle source be stepped down to a 24-volt, 60-cycle source using the Adams Rite 4605 transformer. Why 24 volts? Two reasons. First, at 24 volts, few safety problems occur. Second, popular low-voltage replacement transformers are readily available at any electrical supply house. (Caution: Make sure any replacement transformer has a minimum rating of 20 volts.)

Q: Should the strike be audible or silent?

A: First, understand what causes the strike to be audible or silent. Alternating current changes direction 60 times per second. The noise you hear in the audible unit is that brief period of time during which the solenoid plunger is released from the pole piece as the current builds back up to a peak. This is why a continuous-duty, alternating current system is not recommended. The solenoid simply beats itself to death. By contrast, dc current flows in one direction and, when the coil is energized, the plunger remains seated until the circuit is broken. Silent operation can be achieved in an ac circuit by specifying that an Adams Rite 4603 rectifier be installed between the transformer secondary and the solenoid coil, which then sees only dc current.

Q: What about current draw?

A: The Adams Rite Electric Strike combines standard components that provide the maximum mechanical force necessary to do the job. To the overwhelming majority of intermittent-duty applications, the current draw for this heavy-duty performance poses no problem. However, there are applications where, because of sensitive components somewhere else in the electrical system, a low maximum amount of current flow can be tolerated. In this case, specific hardware is used.

You can use specially wound coils to compromise the mechanical force range through which the strike gives peak performance. This means substituting a solenoid that meets the current draw requirements, but doesn't have the strong plunger "pull" of the high-current solenoid. In most instances, customers' requirements are satisfactorily met with standard available parts without the added expense of specially wound coils.

Q: What is surge current?

A: *Surge current* is the momentary high draw necessary to start the plunger moving from its rest position. Considerably less current is required to hold the plunger, the same way it requires more energy to start a freight train moving than it does to keep it moving.

Q: What is a current limiter?

A: A *current limiter* is simply a combination of known electrical components wired in such a way as to provide the extra surge of current without damaging effects on other equipment in the circuit. One such combination uses capacitors to store electrical energy in exactly the same way an air reservoir stores pneumatic energy, diodes, resistors of correct value, and switches.

This combination stores electrical energy and, upon demand (closing a switch), provides the solenoid coil extra momentary charge. It does so outside the circuit, so the circuit sees only the current required to hold the coil in position.

Q: A 7100 series electric strike has been installed, but is not releasing properly. When the unit is taken out of the jamb and held upside down, it seems to function normally. Are the 7100 series electric strikes handed or what could be the problem?

A: All 7100 series electric strikes are nonhanded. If an electric strike functions correctly in one hand and not the other, this is generally a sign that the strike is being underpowered. An example would be an electric strike being rated at 24VDC, but the power being supplied is 24VAC. The strike will probably operate in one hand and not the other simply because gravity is helping the plunger and blocking arm fall to the unlocked position. The first step would be to verify what voltage the strike is rated for. The solenoid data should be labeled on the side of the strike for easy reference. Make sure the rated voltage of the power source is within ±10 percent of the rated voltage for the solenoid. An incorrect voltage source can and will cause problems with the strike or the entire system.

Q: Currently installed on a single aluminum door and frame is an MS1850S deadbolt. The customer is requesting that an electric strike be tied into an access control system. What hardware is needed?

A: The first step would be to replace the MS1850S with a 4900 deadlatch of the same backset. The 4900 will fit into the existing preparation for the MS1850S deadbolt without any modification. The 7130 electric strike is required to cover the strike opening or cutout in the frame. A 4900 paddle or any style of Adams Rite latch handle is needed for the interior side of the door. The existing mortise cylinder on the exterior of the door can be used on the 4900 deadlatch without any changes. Other required components include some type of control switch (for example, a pushbutton, a keypad, or a card reader) and power source (a transformer and possibly a rectifier) for the strike.

Troubleshooting

Accurate checking of an electrical circuit requires the proper tool. Purchase a good 200,000-volt-per-ohm volt-ohm-milliammeter (VOM). Simpson Model 261, one of several on the market, costs about $130. Read the instructions and make some practice runs on simple low-voltage circuits.

Checking Voltage

1. Zero the pointer.
2. Turn off the power to the circuit being measured.
3. Set the function switch to the correct voltage to be measured (+dc or ac).
4. Plug the black test lead into the common (−) jack. Plug the red test lead into the (+) jack.
5. Set the range selector to the proper voltage scale. *Caution:* It is important that the selector be positioned to the nearest scale above the voltage to be measured.
6. Connect the black test lead to the negative side of the circuit and the red lead to the positive side. This is applicable to dc circuit only. Turn power on to the circuit to be tested. If the pointer on the VOM moves to the left, the polarity is wrong. Turn the switch function to −dc and turn the power on again. The pointer should now swing to the right for proper reading on the dc scale.
7. Turn off the circuit before disconnecting the VOM.

Note: For an ac circuit, connections are the same, except you don't have to worry about polarity.

Checking DC Current

1. Zero the pointer.
2. Turn off the power to the circuit being measured.
3. Connect the black test lead to the −10-amp jack and the red test lead to the +10-amp jack.
4. Set the range selector to 10 amps.
5. Open the circuit to be measured by, for example, disconnecting the wire that goes to one side of the solenoid. Connect the meter in series. Hook the black lead to one of the disconnected wires and the red lead to the other wire.
6. Turn the power on to the circuit and observe the meter. If the pointer moves to the left, reverse the leads in the −10-amp and +10-amp jacks.
7. Turn the power back on to the circuit and read the amperage on the dc scale.
8. Turn the power off before disconnecting the VOM.

Checking Line Drop

Measure line drop by comparing voltage readings at the source (the transformer's secondary or output side) with the reading at the strike connection.

Locating Shorts

The VOM is the most reliable instrument for detecting a short. This is accomplished by setting up the VOM to measure resistance.

1. Set the range switch at position Rxl.
2. Set the function switch at +dc.
3. Connect the black test lead to the common negative jack and the red test lead to the positive jack.
4. Zero the pointer by shorting test leads together.

5. Connect the other ends of the test lead across the resistance to be measured. For a solenoid, connect one end of the test lead to one coil terminal and the other to the other terminal.

Watch the meter. If the pointer does not move, the resistance being measured is open. If the pointer moves to the peg on the right-hand side of the scale, the resistance being measured is shorted closed. If you get a reading in between these two extremes, there is probably no problem with the solenoid.

The Strike Will Not Activate After Installation

1. Check the fuse or circuit breaker supplying system.
2. Check that all wiring connections are securely made. When wire nuts are used, be sure both wires are twisted together.
3. Check the solenoid coil-rated voltage (as shown on the coil label) to make sure it corresponds to the output side of the transformer within ±10 percent.
4. Using the VOM, check the voltage at the secondary (output) side of the transformer.
5. Using the VOM, check the voltage at the solenoid to make sure there are no broken wires, bad rectifiers, or bad connections.
6. Check the coil for a short.

The Strike Will Not Activate After Use

1. Check the fuse or circuit breaker supplying system.
2. Make sure you have a transformer. One shot of 115 ac volts will ruin the coil.
3. Make sure the rated voltage of the transformer and the rated voltage of the coil correspond with ±10 percent.
4. Check the coil for a short.

Overheating

If the transformer overheats, proceed as follows.

1. Make sure the rated voltage of the transformer and the rated voltage of the coil correspond within ±10 percent.

2. Make sure the volt-amp rating is adequate. We recommend a minimum of 20 to 40 volts.

If the rectifier overheats, proceed as follows.

1. The rectifier might be wired wrong, which means the overheating is a temporary situation because it burns out.

2. There might also be too many solenoids supplied by a single rectifier, and more current is being pulled through than the diodes can handle.

If the solenoid overheats, proceed as follows.

1. First, define overheating. The coils used by Adams Rite have a temperature rise rating of 65°C (149°F) above ambient. However, all 7800 series continuous-duty units should run 200°F or less in a 72°F environment.

2. The vast majority of these intermittent-duty units never see the kind of use that brings the coil to maximum rating. If a coil gets extremely hot on very short pulses at two- or three-second intervals, either the wrong coil or the wrong transformer output is being used. The same is basically true for continuous-duty coils. If the coil temperature exceeds the ratings, it must be because the coil voltage or the transformer is improperly coordinated.

3. Set the meter up as if testing for a short and obtain the exact resistance.

The rest of this chapter is a reprint of the Electric Strike Handbook, courtesy of Adams Rite Manufacturing.

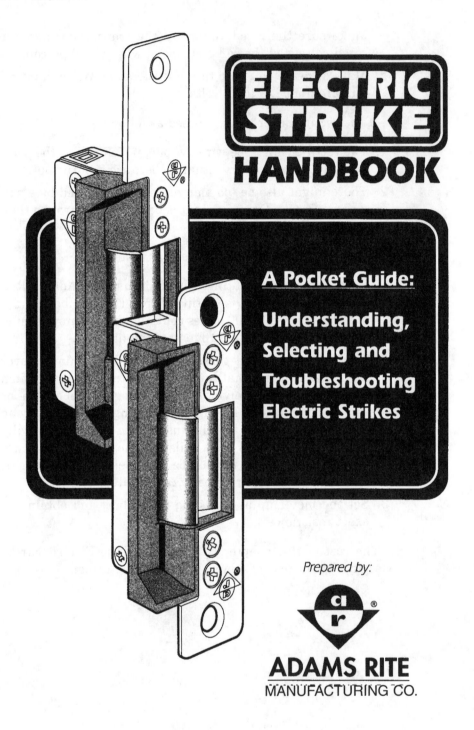

ELECTRIC STRIKE HANDBOOK

A Pocket Guide:

Understanding,
Selecting and
Troubleshooting
Electric Strikes

Prepared by:

ADAMS RITE
MANUFACTURING CO.

TABLE OF CONTENTS

INTRODUCTION

This handbook was designed as both an educational guide and as a quick reference tool. It offers a broad understanding of electric strikes and provides answers to common questions associated with their installation and usage.

Electric strikes are not overly complicated devices, but there are many factors to consider when purchasing and installing them. Attention to the Strike Selection portion of this handbook is, of course, the best way to avoid problems listed in the Troubleshooting section.

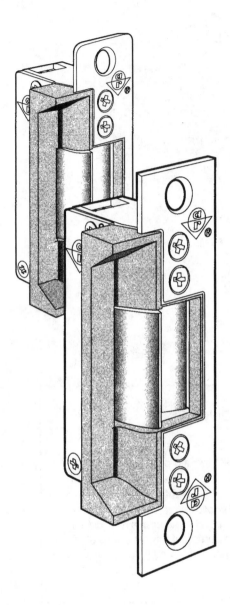

STRIKE SELECTION

Electric strikes are architectural hardware devices that provide remote activation and add security features to a traffic control device. They are highly "pick" and "shake" resistant and can have a long, maintenance-free life when properly installed and powered.

The fact that electric power is used for its operation does not make it an electrical appliance. With a few exceptions, the electrical circuit is designed to meet the needs of the strike, not the other way around. The hardware specifier should select the strike and expect the electrician to provide the power at the point of installation. In cases where an existing circuit is to be used, a strike must be selected to match it.

HARDWARE CONSIDERATIONS
There are nine considerations to be covered:

Face Shape: All Adams Rite electric strikes have flat faces, except for the 7101 and 7131, which have radiused faces to match the nose shape of paired narrow stile glass doors.

Face Size: The basic Adams Rite size conforms to American National Standards Institute (ANSI) strike preparation: 1-1/4" x 4-7/8". However, other sizes are offered to fill or cover existing jamb (or opposing stile) preparation from a previous installation.

Face Corners: Strikes are available with ROUND corners for installation in aluminum (where preparation is usually done by router) and with SQUARE corners for punched hollow metal ANSI preparation or wood mortise.

Jamb Material: Adams Rite has strike models suitable for aluminum, steel or wood jambs. They are designed to fit jamb sections as shallow as 1-3/16". (See Application Chart on page 15).

Previous Strike Preparation: If jamb was previously fitted with a 7810 strike, the new 7110 unit will fit with no alteration. In the case where a hollow jamb was originally prepared to receive the bolt from an Adams Rite MS® deadlock and a 4510 or 4710 latch is to be substituted, the 7130 (flat jamb) or 7131 (radius inactive door) will cover the old strike cutout slot.

Lip Length: The standard lip on all basic Adams Rite electric strikes accommodates a 1-3/4" thick door which closes flush with the jamb. Where door/jamb relationship is different, a lip extension can be added.

Compatible Latch: Adams Rite has strike models that are compatible with other manufacturers' key-in-knob (bored locks) or mortise latches and with our own 4500 or 4700 Series deadlatches, and mortise or rim type exit devices. (See Application Chart on page 15).

Finish: Standard anodized finishes for most models are: Satin Aluminum (628), Dark Bronze (313) and Black (335). See Ordering Matrix on page 15 for other finishes available on some models.

Handing: All standard operation strikes are unhanded. They can be installed for either right or left hand doors.

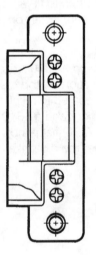
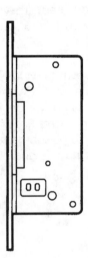
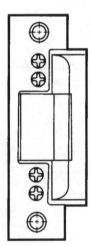

HARDWARE CONSIDERATIONS
Installing in different materials

If an electric strike is installed in a wood or aluminum jamb, the required shape of the jamb cut-out is usually not much of a concern. The reasons for this are that wood is easy to work and/or rework and aluminum is routed onsite. In addition, the door frames are usually mortised for the strike by the installing carpenter after the frame installation is completed.

When installing an electric strike into a steel (hollow metal) frame, it is important to keep in mind the ANSI standard that was developed to determine size, shape and location for cutouts that must be made on the door and frame. The standard was developed because steel frames are pre-punched during manufacturing to accept various hardware items.

The shape of the strike cut-out is defined by ANSI for flat metal strike plates used with every lockset. This cut-out is 4-7/8" x 1-1/4" x 3/32" deep, with a 3-3/8" x 3/32" thick strike lip. This shape is the same for both mortise locks and cylindrical (key-in-knob) locks.

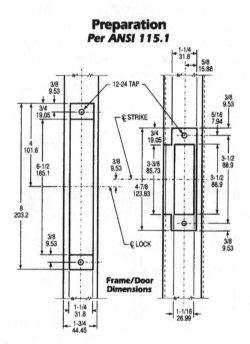

Preparation
Per ANSI 115.1

Frame/Door Dimensions

Installing for different lock types

The requirements for installing an electric strike are most often not determined until after the door and frame are installed, due to the fact that final foot traffic patterns are not finalized until the building is complete. The result is that the electric strikes have to be hand cut into the door frame.

If the door is using a cylindrical-type lock, installing the electric strike is relatively easy. ANSI 161.1 requires that the centerline of the lock's latchbolt and the centerline of the strike be the same distance above the floor. Since most electric strikes are built with the bolt retainer jaw in the center of the faceplate, the door frame face will require only minor modification to permit the installation of an electric strike.

If the door uses a mortise lock, ANSI 115.1 requires that the centerline of the lock's cut-out in the door be 3/8" below the centerline of the strike's location in the frame. Further complicating the situation is the fact that each mortise lock manufacturer positions the latchbolt at a unique location on the lock edge. The reason for the complication in ANSI 115 stems from the fact that mortise locks pre-date the development of ANSI standards. While Adams Rite electric strikes are compatible with latches from virtually all top manufacturers (provided that the strike can be located on center of latchbolt), for retrofit applications, the 7900 Series (or similar offset strike) may be the only answer.

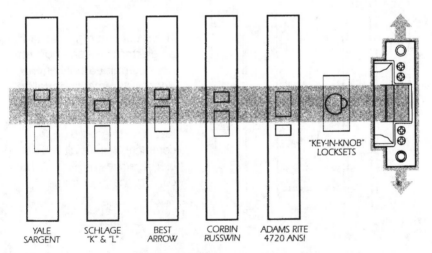

YALE SARGENT   SCHLAGE "K" & "L"   BEST ARROW   CORBIN RUSSWIN   ADAMS RITE 4720 ANSI

"KEY-IN-KNOB" LOCKSETS

This diagram shows how different lock manufacturers position the latchbolt at unique locations on the lock's edge. With proper location, Adams Rite electric strikes are compatible with latches from virtually all top manufacturers.

ELECTRICAL CONSIDERATIONS

The first electrical factor to determine is DUTY. Is the operation to be intermittent or continuous? If the door is normally locked and released only momentarily from time to time, then it is intermittent duty. If it is the relatively rare case where the strike is activated (unlocked) for long periods, the duty is continuous. A still rarer requirement is continuous ("fail-safe") in which the strike is locked only when its current is switched on. Adams Rite 7100 Series Strikes are field convertible from one mode to another. However, an AC intermittent solenoid must not be used continuously in either.

Intermittent Duty:	For a normal intermittent application, specify an electric strike using 24V AC. This gives enough power for almost any entrance, yet this low voltage range is below that requiring U.L. or Building Code supervision. Good, reliable transformers (4605/4606) are available in this voltage.
	The buzzing sound inherent in AC activated strikes is usually not considered offensive in intermittent use. In fact, it acts as an audible signal to a person waiting to enter. If a silent operation is desired, specify 24 VDC and add a 4603 Rectifier to the AC circuit. The rectifier changes 24 VAC to 24 VDC. DC current is silent while operating the strike.
Continuous Duty:	Continuous duty is required when the strike will be energized for more than about 60 seconds at a time. Most continuous duty applications can be supplied through the same type 115/24 VAC transformer used for intermittent jobs. Specify "AC Continuous" and Adams Rite will automatically add the rectifier necessary to achieve continuous performance on the 24 VAC voltage specified.
Fail-Secure vs. Fail-Safe	Fail-Secure strikes are by far the most popular configuration. A fail-secure strike, with no power applied, is locked or secure. Power is then applied to release the blocking mechanism so that the door can be pulled open. Fail-Safe is the opposite. With no power applied, the strike is safe or unlocked. Fail-Safe strikes are generally energized for long periods.

The main question to ask when selecting this feature is what condition the strike needs to be in if power were to fail. A fail-secure strike remains or reverts to a locked condition, while the fail-safe unlocks. For this reason a fail-safe strike, in most cases, would not be used on the main entrance of a business. Security of the building is compromised in the event of a power failure.

Fail-safe strikes are many times ordered incorrectly. The misconception is that a fail-secure strike will trap someone inside a building. The fact is that, from the inside, an operator such as a lever or pushbar retracts the latch to allow for exiting. Only in specialized applications should a fail-safe strike be used. In addition, a fail-safe strike should never be used in a fire-rated application. The design and control of the circuit is affected by which configuration is desired.

Monitoring: If a visual or other signal is required to tell the operator the electric strike status, a "monitoring" strike is needed. Two sensor/switches are added: one is activated by the latchbolt's penetration of the strike and the other by the solenoid plunger that blocks the strike's release. The switches can also be used for "mantrap" applications, where power to one strike is controlled by the switch(es) of the other.

Transformer: Low voltage for electric strike operation is obtained by the use of a tranformer, which steps down the normal 115 volt AC power to 12 AC, 16 AC or 24 AC volts. The electrical specifier needs to call out three items: (1) Input voltage (usually 115 VAC) ; (2) Output voltage (12, 16 or 24 VAC); (3) Capacity of transformer, called volt-amps (Output voltage X output amps). Skimping on the capacity of the transformer to save a few dollars will generally under-power the door release and is likely to bring complaints of "poor hardware."

The Wire: The wire must carry the electrical power from the transformer through the actuating switch (or switches) to the door release. It must be large enough to minimize "frictional" line losses and deliver most of the output from the transformer to the door release. Just as a small diameter garden hose won't provide a full flow of water from the nozzle, particularly if it's a long run, neither will an under-sized wire carry the full current. (See chart and formula for determining wire size on page 14).

Electrical Trouble-shooting: When insufficient electrical power is suspected in a "weak" door release, a simple check can be made. Measure the voltage at the door release while the unit is activated. If the voltage is below that specified on the transformer, the problem is in the circuit, probably an under-capacity tranformer if the circuit length is short. A long run may be both a transformer and wire size problem.

SOME QUESTIONS AND ANSWERS ABOUT ELECTRICAL CONSIDERATIONS

What is the electrical voltage source?

The majority of electric strikes are used where 115 Volt AC 60 cycle alternating current is the power source. Unless there are specific customer demands to do otherwise, the 115 VAC 60 cycle source should be stepped down to 24 VAC 60 cycle by use of the Adams Rite 4605/4606 Transformer. This is recommended because at 24 volts there is very little safety problem. Also, popular low voltage transformers are available at any electrical supply house. (Caution: Make sure any replacement transformer matches the circuit requirements.)

Audible or Silent?

Alternating current changes direction 60 times per second. The audible "buzz" in the unit is that brief instant when the solenoid plunger is released from the pole piece in the coil as the current

decreases through zero and then pulled back against the pole piece as the current builds back up to a peak. Thus, continuous duty alternating current systems are not recommended. By contrast, DC current flows in one direction and when the coil is energized the plunger remains seated until the circuit is broken. Silent operation of a DC strike is achieved in an AC circuit by specifying an Adams Rite 4603 Rectifier to be installed between the transformer secondary and the solenoid coil, which then sees only DC current.

Can more than one strike be on a given circuit?

Most circuits involve a single strike powered by a single power source. When multiple strikes (loads) are to be wired in a circuit the strikes must be connected in parallel. Having a dedicated power source for each strike has advantages in that if a power source fails, it affects only one opening. Having one power source supply multiple strikes affects all the openings within the circuit.

What are Suppression Devices?

When an inductive load (solenoid) is deenergized, the electromagnetic field collapses, sending voltage back through the circuit. This voltage, over time, can damage control switch or relay contacts. Electrical components can be arranged to limit the effects of this inductive "kickback." Having the rectifier placed between the load and the switch will suppress the kickback. A diode placed in parallel with the switch will protect the contacts in a DC circuit. In AC or DC circuits, use a Metal Oxide Varistor (MOV) to protect the switch contacts in this manner.

What is "Inrush" current?

Inrush current is the momentary high draw necessary to start the plunger to move from its rest position. Considerably less current is required to "hold" the plunger, in the same way that it requires more energy to start a freight train moving than it does to keep it moving. Consider inrush current when selecting components within a circuit.

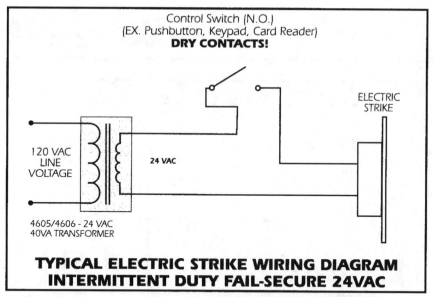

Control Switch (N.O.)
(EX. Pushbutton, Keypad, Card Reader)
DRY CONTACTS!

ELECTRIC
STRIKE

120 VAC
LINE
VOLTAGE

24 VAC

4605/4606 - 24 VAC
40VA TRANSFORMER

TYPICAL ELECTRIC STRIKE WIRING DIAGRAM
INTERMITTENT DUTY FAIL-SECURE 24VAC

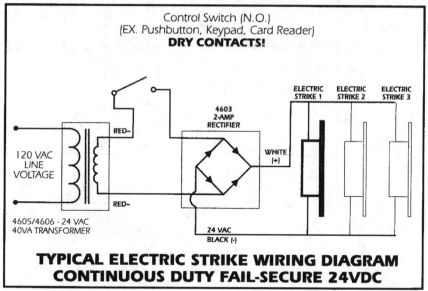

Control Switch (N.O.)
(EX. Pushbutton, Keypad, Card Reader)
DRY CONTACTS!

ELECTRIC   ELECTRIC   ELECTRIC
STRIKE 1   STRIKE 2   STRIKE 3

4603
2-AMP
RECTIFIER

RED~

120 VAC
LINE
VOLTAGE

WHITE
(+)

RED~

4605/4606 - 24 VAC
40VA TRANSFORMER

24 VAC
BLACK (-)

TYPICAL ELECTRIC STRIKE WIRING DIAGRAM
CONTINUOUS DUTY FAIL-SECURE 24VDC

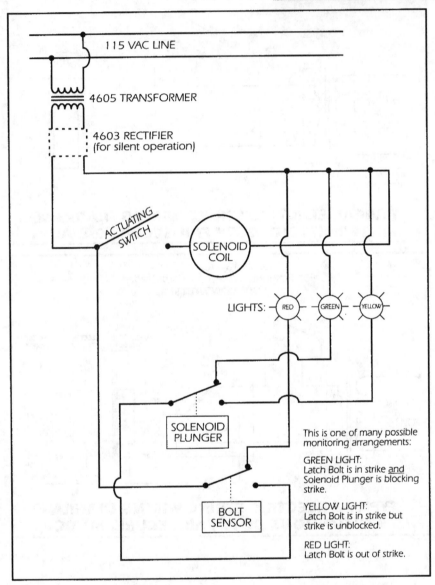

Sample Wiring Diagram for Monitor Signal Operation

Wire Selection

In practice, 18-ga stranded should be the minimum size wire used for system wiring. In applications with long runs of wire, or when operating units that have high current requirements, use the chart below as a guide for selecting the proper wire size. Stranded wire, versus solid wire, is recommended for electrified hardware applications. Stranded wire offers greater resistance from the vibration and flexing within door and framing systems. For more information concerning wiring of low-voltage circuits refer to the National Electrical Code (NFPA 70).

TOTAL LENGTH OF WIRE RUN ONE DIRECTION (FT.)	CURRENT at 24V			
	.5 Amp	1 Amp	1.5 Amp	2 Amp
0-100	18	18	16	14
100-150	18	16	14	12
150-200	18	16	12	12
200-300	18	14	12	10
300-400	18	12	10	–

TOTAL LENGTH OF WIRE RUN ONE DIRECTION (FT.)	CURRENT at 12V			
	.5 Amp	1 Amp	1.5 Amp	2 Amp
0-100	18	14	12	12
100-150	16	12	12	10
150-200	14	12	10	–
200-300	12	10	–	–
300-400	12	–	–	–

Solenoid Data

Description	Lead Color	Coil Resistance (OHMS ± 5%)	Peak Instantaneous Current (AMPS)	Continuous or Hold Current (AMPS)	Peak Instantaneous Power (WATTS)	Continuous or Hold Power (WATTS)
24 VDC Cont.	White	141.6	.170	.170	4.09	4.09
16 VDC Cont.	Green	61.8	.222	.222	3.05	3.05
12 VDC Cont.	Red	34.6	.332	.332	3.81	3.81
24 VAC Int.	Red	34.6	.744	.431	19.15	6.43
16 VAC Int.	Blue	16.3	1.030	.636	17.30	6.60
12 VAC Int.	Yellow	8.8	1.420	.813	17.74	5.82

Ordering Matrix

The Adams Rite
Numbering System
consists of six factors
that make up the final
Part Number.

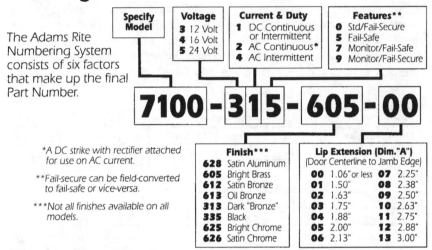

Specify Model	Voltage	Current & Duty	Features**
	3 12 Volt	**1** DC Continuous or Intermittent	**0** Std/Fail-Secure
	4 16 Volt	**2** AC Continuous*	**5** Fail-Safe
	5 24 Volt	**4** AC Intermittent	**7** Monitor/Fail-Safe
			9 Monitor/Fail-Secure

7100 - **315** - **605** - **00**

*A DC strike with rectifier attached for use on AC current.

**Fail-secure can be field-converted to fail-safe or vice-versa.

***Not all finishes available on all models.

Finish***		Lip Extension (Dim."A")	
		(Door Centerline to Jamb Edge)	
628	Satin Aluminum	**00** 1.06"or less	**07** 2.25"
605	Bright Brass	**01** 1.50"	**08** 2.38"
612	Satin Bronze	**02** 1.63"	**09** 2.50"
613	Oil Bronze	**03** 1.75"	**10** 2.63"
313	Dark "Bronze"	**04** 1.88"	**11** 2.75"
335	Black	**05** 2.00"	**12** 2.88"
625	Bright Chrome	**06** 2.13"	**13** 3.00"
626	Satin Chrome		

Application

MATERIAL LATCH TYPE	ALUMINUM JAMB	HOLLOW METAL JAMB	WOOD JAMB	ALUMINUM STILE
Adams Rite Deadlatch	**7100** **7110/11** **7130**	**7140**	**7110/11** **7140**	**7101** **7131**
Adams Rite Mortise Exit Device	**7108**	**7128**	**7128**	
Key-in-Knob or Lever	**7100** **7110/11**	**7140** **7240**	**7110/11** **7140**	
Mortise Lock (without Deadbolt)	**7160**	**7170** **7270** **7900**	**7170**	
Rim Panic Device (by Others)	**71R1**	**71R1**	**71R1**	

The chart above summarizes which strike models are suited for different applications.

How To Check:

Accurate checking of an electrical circuit requires the proper tool, such as a good 20,000 volt per ohm meter (VOM) or Digital Multimeter (DMM). Of the two types of meters available, the DMM offers greater accuracy and ease of use. The auto-ranging features of a DMM, along with the ability to read both positive and negative voltages, helps even the novice take measurements like a pro. Read the instructions and make some practice runs on simple low voltage circuits.

How to check Voltage:

(a) Zero the pointer (if analog type meter).

(b) Be sure power is turned off to the circuit being measured.

(c) Set the function switch to the correct voltage to be measured (+DC or AC).

(d) Plug black test lead into the common (–) jack. Plug the red test lead into the (+) jack.

(e) If required, set the range selector to the proper voltage scale. Caution: It is important that the selector be positioned to the nearest scale above the voltage to be measured.

(f) Connect black test lead to the negative side of the circuit and the red lead to the positive side. This is applicable to DC circuit only. Turn power on to the circuit being tested. Turn switch function to DC and turn power back on. Pointer should now move to right for proper reading on the DC scale. AC circuit connections are the same except as noted in (c) above. It should also be noted that polarity in the AC circuit is not a concern.

(g) Be sure to turn the circuit "OFF" before disconnecting multimeter.

How to check DC Current:

(a) Zero the pointer (if analog type).

(b) Be sure power is turned off to the circuit being measured.

(c) Connect black test lead to the -10A jack and the red test lead to the +10A jack.

(d) Set range selector to 10 AMPS.

(e) Open the circuit to be measured by disconnecting the wire that goes to one side of the solenoid. Connect the meter in series by hooking the black lead to one of the disconnected wires and the red lead to the other wire.

(f) Turn the power on to the circuit and observe the meter.

(g) Be sure and turn "OFF" before disconnecting multimeter.

How to find Shorts:

The multimeter is the most reliable instrument for detecting a short. This is accomplished by setting up the multimeter to measure resistance.

(a) Set the range switch at position R x1.

(b) Set function switch at +DC.

(c) Connect black test lead to the common (−) jack and red test lead to the (+) jack.

(d) Connect the other ends of the test lead across the resistance to be measured. In the case of a solenoid, connect one end of the test lead to one coil terminal, the other end of the test lead to the other terminal.

(e) Watch meter. If there is no movement of the pointer, the resistance being measured is OPEN. If the pointer moves to the peg on the right hand side of the scale, the resistance being measured is CLOSED. If it reads in between these two extremes, it's very likely the solenoid is OK. Compare the measured value to the solenoid data chart on page 14.

What To Do When:

Strike will not activate after installation:

(a) Check fuse or circuit breaker supplying system.

(b) Check to make sure all wiring connections are securely made. When wire nuts are used, care must be taken to be sure both wires are twisted together.

(c) Check the solenoid coil rated voltage (as shown on coil label) to make sure it corresponds to the output side of the transformer within ±10%.

(d) Using a multimeter, check the voltage at the secondary (output) side of the transformer.

(e) Using a multimeter, check the voltage at the solenoid. This will ensure there are no broken wires, bad rectifiers or bad connections.

(f) Check coil for short.

Strike will not activate after use:

(a) Check fuse or circuit breaker supplying system.

(b) Make sure you have a transformer. One shot of 115 VAC and the coil is gone.

(c) Make sure the rated voltage of the transformer and the rated voltage of the coil correspond within ±10%.

(d) Check coil for short.

Transformer overheats:

(a) Make sure the rated voltage of the transformer and the rated voltage of the coil correspond within ±10%.

(b) Make sure the VA rating is adequate. 40 VA is recommended, with 20 VA being the absolute minimum. Transformer heating may be experienced even in moderate use applications.

Rectifier overheats:

(a) The rectifier is wired wrong, which means the over-heating is a temporary situation (a few milliseconds) and then it's burned out, or

(b) There are too many solenoids being supplied by a single rectifier and more current is being pulled through it than the diodes are rated for.

Solenoid overheats:

(a) The coils used by Adams Rite have coil temperatures that range from 130° to 200°. Regardless of the exact degree, this is something too hot to touch.

(b) Check the solenoid coil rated voltage (as shown on coil label) to make sure it corresponds to the output side of the transformer within ±10%.

(c) The vast majority of Adams Rite intermittent duty units never see the kind of use that brings the coil to maximum rating. If a coil gets extremely hot on very short pulses at two or three second intervals, either the coil is wrong or the transformer output is wrong. The same is true for continuous duty coils. If the coil temperature exceeds the ratings, it is because the coil voltage or the transformer are improperly matched.

(d) If the multimeter is set up to test for a short and the exact resistance is obtained, the results can be compared with specifications found on page 14 to know if the coil is correct.

Sluggish operation:

Verify voltage being supplied to the strike. When operation is sluggish or if the strike operates fine in one hand and not the other, it is typically a sign of AC power being supplied to a DC electric strike.

Strike buzzes:

(a) When using AC intermittent strikes, this is a normal indication, as AC intermittent strikes are designed to buzz to indicate that the door is open.

(b) If supplying AC power to a DC electric strike, a rectifier must be used to convert the current.

(c) Do not use a half wave rectifier. A full wave bridge rectifier is required to ensure silent operation.

Door does not latch easily:

If the door does not latch easily or does not release properly, check for a tight door gap. The strikes are designed to operate with an industry standard 1/8" door gap. Check also for improper fit of door which can cause a pre-load on the strike jaw.

GLOSSARY
ELECTRICAL TERMS

Voltage (Volt): The electric potential energy that causes current to flow in a conductor. Characterized by the symbol E or V. Analogous to the pressure in the water pipes in a home. (See Chart I).

Ampere (Amp): Unit of measurement of the rate of electrical current flow. Characterized by the symbol I or A. Analogous to gallons of water flowing past a given point.

Direct Current (DC): This is the term used to describe the flow of electrical current in one direction only, uniform and continuous in the conductor. (See Chart I).

Alternating Current (AC): Differs from direct current in only one very important point. It typically starts at zero, gradually increases to a maximum, then gradually decreases to zero. This change is then repeated in the opposite direction. Alternating Current flows back and forth in the conductor as a pendulum swings back and forth. (See Chart I).

Resistance: The opposite of current flow in a conductor. It is analogous to friction in a mechanism. The unit of electrical resistance is the OHM and is characterized by the symbol W or R.

Watt: A unit of electrical power. Power is work being accomplished. It is characterized most often by the letter P. Power in a DC circuit is the product of the voltage multiplied by the amperage. In an alternating circuit and for resistive loads such as the solenoid in electric strikes, it is the product of the effective voltage multiplied by effective current. The word *effective* means the amount of alternating current that produces the same heating effect that a given amount of direct current produces. For non-resistive loads, such as for a motor, it is the product of the effective current multiplied by the power factor.

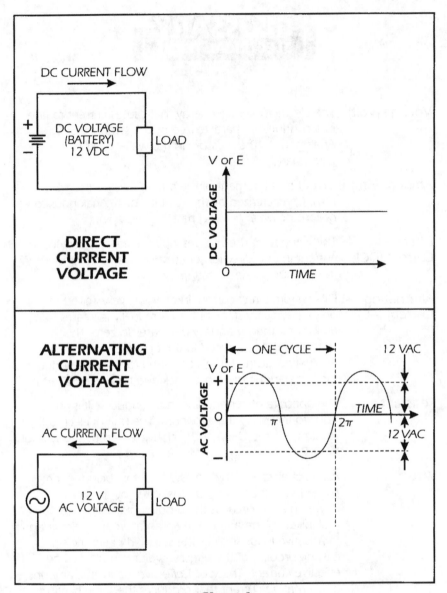

Chart I
AC/DC Voltage

Load:	Any object that consumes electrical power, such as a light bulb or electric motor. In some applications the load is a solenoid.
Cycle:	Also referred to as Frequency – Hertz. The number of times, per second, the current in an alternating current system reverses its direction of flow. 60 cycle is most common to the U.S., however, 25 cycle and 50 cycle are also in use. (See Chart I).
Duty Cycle:	Refers to the length of time and the frequency any electrical device is started and stopped. Continuous duty simply means the electrical device is energized (started) and left that way for an indefinite period of time. Intermittent means the device is energized for short periods of time and then turned off for minimum periods of time. Example: On two seconds, off ten seconds, etc.
Volt-Amp (VA Rating):	The product of rated input voltage multiplied by the rated current. This establishes the "apparent energy" available to accomplish work.
Line Drop (Voltage Drop or Potential Difference):	Any electrical circuit experiences "line drop" resulting from two principal factors: (1) size of the wire – the smaller the wire used, the higher the "frictional" type resistance to the current flow. Thus, if readings were taken at various intervals "downstream" of the power source, a progressive drop in voltage would be observed. This fact can be somewhat offset by proper sizing of wire to the given voltage and length of run; (2) length of run (the distance the electrical device is from its source of power). Obviously, the farther out on the line the device to be electrically operated is, the less power there is available to do the job. See Wire Selection chart on page 14.

GLOSSARY
ELECTRICAL TERMS

Circuit:
The path through which the electrical energy flows to and from the source to the device being operated.

Conductor:
What the trade commonly calls "electric wiring". A more accurate definition of the word describes the ability of any given material to carry electrical current. Examples of materials that are good conductors and offer little resistance to the flow of electricity are copper, silver, aluminum and carbon graphite, to mention a few. Copper is the most common material used in wiring because it combines low resistance to current flow at reasonable cost. Poor conductors are such materials as green wood, distilled water, moist earth. As a material goes beyond the poor conductor classification, it enters the insulator classification. Insulators have such high resistance to the flow of electricity that none can pass. Typical of those materials are glass, ceramic, mica, rubber and some plastics.

Transformer:
A transformer is a device for transferring energy in an alternating current system from one circuit to another. Thus, energy at high voltage may be transformed to energy at low voltage and vice-versa. There are many kinds of transformers, but electric strike applications basically require a good quality, step down transformer. A 40 VA rating is recommended. For these applications the high voltage input side of a transformer is called the "Primary" side. The output or low voltage side of the transformer is called the "Secondary."

Switch:
A switch can be described as a device placed in the electrical circuit in order to "make" or "break" the flow of electrical current.

Rectifier: An electrical unit designed to convert alternating current to pulsating direct current. This is accomplished by use of diodes which are the electrical equivalent of a water system check valve, permitting flow in one direction only. (See Chart II). Full-wave rectification is required for proper strike operation.

OHM's Law: One of the most widely used principles of electricity. It expresses the ratio of voltage (E), current (I) and resistance (R). The following equations are used in calculating these ratios:

$$E = I \times R \quad I = \frac{E}{R} \quad R = \frac{E}{I}$$

Ground: Connects the electrical system to the earth. The third (round) prong on an electrical cord is a ground plug. The green wire with spade terminal on an adapter plug (which is supposed to be connected to cover plate mounting screw) is also a ground. The purpose is to minimize danger from shock and prevent lightning from "running-in" and destroying the electrical system.

Short: Improper connection between "hot" current carrying wire and neutral or ground.

Polarity: Very broadly, it expresses the electrical phenomenon in which like poles (North-North) repel and opposites (North-South) attract. In a DC circuit, it describes the direction electrons are flowing.

Solenoid & Plunger: The solenoid itself is a coil of copper windings which are insulated from each other. The magnetic pull strength and current draw are determined by the number of windings. The plunger is a bar (sometimes laminated) of soft iron or steel. This plunger becomes magnetized when the coil is energized and the resulting plunger movement can do mechanical work.

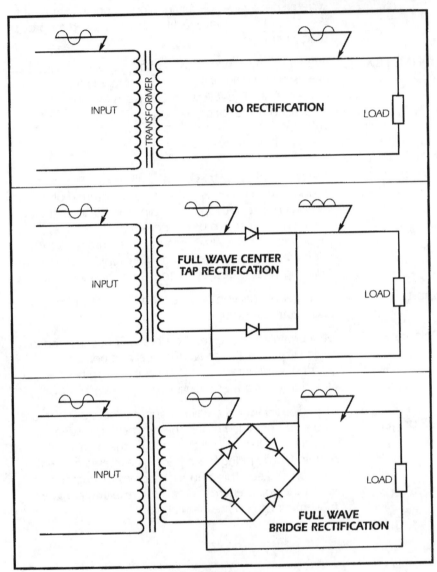

Chart II
Rectifier Circuits

11

Intruder Alarms

This chapter gives an overview of wireless and hardwired alarm system components, and it explains how to install them.

How much of a deterrent intruder alarms are is hard to say, because people who use them may be more security conscious than people who don't. Perhaps alarm users as a group have better doors, windows, and locks than do people without alarms. But some anecdotal evidence exists that alarms are useful.

More than 600 inmates in an Ohio prison were asked what single thing they would want to protect their homes from burglars. The most popular choice was a dog; the next, a burglar alarm. In a Security Industry Association survey, 85 percent of the 1,000 police officers interviewed said they believe security systems decrease the chances of a home being burglarized.

In movies, super spies and master criminals often defeat the most sophisticated security systems. Those systems include such items as retina scanners, fingerprint readers, closed-circuit television systems, and strategically placed ultrasensitive body heat, movement, and sound sensors.

In the movie *Entrapment*, for example, while wearing a tight leather bodysuit, Catherine Zeta-Jones shimmied along the floor and used sensual dancelike movements to maneuver around an invisible web of protective beams to bypass a state-of-the-art security system. Other movies and television shows later used a similar scene. The fact that you can't defeat an even halfway decent system that way doesn't make the scene less exciting to watch. But, if you install a security system that someone can dance past, you better have good liability insurance.

Modern alarm systems (even low-cost ones) are quite reliable if they're installed right. They can be as sensitive as you want them to be, and they seldom do anything for no reason. So-called "false alarms" are usually the result of a system doing what it was designed to do. How intelligently a system reacts depends on how well you choose and install its various components, and how your customer uses it. If you make bad selection choices, or if your customer doesn't properly use the system, the system will seem to overreact (give false alarms) or overlook an intruder.

Let your customer know that blaming the alarm system for a false alarm is like blaming a trained dog for sitting when you accidentally say "sit," when you mean to say "roll over." That can help your customer appreciate the importance of spending a little time with you to learn how to properly use the system.

Your job is to provide a system the end user can easily operate. No matter how technologically advanced a system may be, if the customer finds it too frustrating to use regularly, it won't be of much value. That's why it's important for you to know about the many available options. If your customer has poor vision, for instance, they might prefer a system that talks or responds to voice commands or, perhaps, has a large backlit English (or other language) screen on the control panel. You also need to consider other special needs, such as whether children or pets are in a home. If you take the time to get to know your customer's needs, instead of taking a "cookie cutter" approach, you can avoid most false alarms and have happy customers.

Types of Intruder Alarms

A basic alarm system consists of a control panel, an annunciator, and at least one sensor (or detection device). The *control panel* (or controller or control box) is the brain of the system. See Figure 11.1 This is where someone arms and disarms the system and programs it to react to various alarm conditions.

The most common *annuciators* in alarm systems are bells, sirens, and strobe lights. Many annunciators used in fire alarm systems combine a siren with a strobe light. Some are made to be installed outdoors.

Figure 11.1   An intruder
alarm control panel
(courtesy of Napco).

Sensors

Sensors are the eyes and ears of the system. *Sensors* sense the
presence of an intruder and relay the information to the sensor's control panel, which then activates the siren, bell, or
siren. Today, you have more detection devices to choose from
than ever before, but if you choose the wrong type or install it
the wrong way, you'll have a lot of false alarms.

Some sensors respond to movement, some to sound, and others to body heat. The principle behind each type is similar. When
an alarm is turned on, the devices sense a "normal" condition.
When someone enters a protected area, the devices sense a dis-
turbance in the normal condition and trigger an alarm.

Most sensors fall within two broad categories: perimeter and
interior. *Perimeter devices* are designed to protect a door, window, or wall. They detect an intruder before entry to a room or
building. The three most common perimeter devices are foil,
magnetic switches, and audio discriminators.

Interior (or *space*) *devices* detect an intruder on entry into a
room or protected area. The five most common interior devices
are ultrasonics, microwaves, passive infrareds, quads, and
dual techs.

Foil

You've probably seen foil on storefront windows. *Foil* is a thin,
metallic, lead-based tape, usually 1/2-inch to 1-inch wide,

that's applied in a continuous run to glass windows and doors. Sometimes, foil is used on walls. Like wire, foil acts as a conductor to make a complete circuit in an alarm system. When the window (or wall or door) breaks, the fragile foil breaks, creating an incomplete circuit and triggering the alarm.

Usually, foil comes in long, adhesive-backed strips and is applied along the perimeter of a sheet of glass. Each end of a run must be connected to the alarm system with connector blocks and wire. Foil is popular in stores because it costs only a few cents per foot, and it alerts would-be burglars that the store has an alarm.

Foil has three major drawbacks:

1. It can be tricky to install properly.

2. It breaks easily when the windows are being washed.

3. Many people consider it unsightly.

Magnetic Switches

The most popular type of perimeter device is the *magnetic switch*, which is used to protect doors and windows that open. Magnetic switches are reliable, inexpensive, and easy to install.

As its name implies, the magnetic switch consists of two small parts: a magnet and a switch. Each part is housed in a matching plastic case. The switch contains two electrical contacts and a metal spring-loaded bar that moves across the contacts when magnetic force is applied. When magnetic force is removed, the bar lifts off one of the contacts, creating an open circuit and triggering an alarm condition.

In a typical installation, the magnet is mounted on a door or a window, and the switch is aligned about 1/2-inch away on the frame. When a protected window or door opens, the magnet and switch move out of alignment, which removes the magnetic force.

Some magnetic switches are rectangular, for surface mounting. Others are cylindrical, for recess mounting in a small hole. The recess-mounted types look nicer and are less conspicuous, but they are a little harder to install.

One problem with some magnetic switches is an intruder can defeat them by using a strong magnet outside a door or window to keep the contacts closed. Some models can be defeated by placing a wire across the terminal screws of the switch and jumping the contacts. Another problem is, if a door is loose fitting, the switch and magnet can move far enough apart to cause false alarms.

"Wide gap" reed switches can be used to solve those problems. Because *reed switches* use a small reed instead of a metal bar, they're less vulnerable to being manipulated by external magnets. The wide gap feature allows a switch to work even if the switch and magnet move from 1 inch to 4 inches apart. Some magnetic switches come with protective plastic covers over their terminal screws. The covers thwart attempts at jumping. Most types of magnetic switches cost only a few dollars each.

Audio Discriminators

Audio discriminators trigger alarms when they sense the sound of breaking glass. The devices are effective and easy to install. According to a survey by *Security Dealer* magazine, over 50 percent of professional alarm installers favor audio discriminators over all other forms of glass-break-in protection.

By strategically placing audio discriminators in a protected area, you can protect several large windows at once. Some models can be mounted on a wall up to 50 feet away from the protected windows. Other models, equipped with an *omnidirectional pickup pattern*, can monitor sounds from all directions and are designed to be mounted on a ceiling for maximum coverage.

A problem with many audio discriminators is they sometimes confuse certain high-pitched sounds—like keys jiggling or a telephone ringing—with the sound of breaking glass, and produce false alarms. Better models require both the sound of breaking glass and shock vibrations to trigger the alarm, which greatly reduces false alarms.

Another problem with audio discriminators is their alarm is triggered by sound. An intruder can bypass the device by cutting a hole through the glass or by forcing a window open. Audio discriminators work best when combined with magnetic switches.

Ultrasonic Detectors

Ultrasonic detectors transmit high-frequency sound waves to sense movement within a protected area. The sound waves, usually at a frequency of over 30,000 cycles per second, are inaudible to humans, but can be annoying to dogs. Some models consist of a transmitter that is separate from the receiver; others combine the two in one housing.

In either type, the sound waves are bounced off the walls, floor, and furniture in a room until the frequency is stabilized. Thereafter, the movement of an intruder causes a change in the waves and triggers the alarm.

One drawback to ultrasonic detectors is they don't work well in rooms with wall-to-wall carpeting and heavy draperies because these soft materials absorb sound.

Another drawback is ultrasonic detectors do a poor job of sensing fast or slow movements and movements behind objects. An intruder can defeat a detector by moving slowly and hiding behind furniture. Although they were popular a few years ago, ultrasonic detectors are a poor choice for most homes today. They can cost over $60; other types of interior devices cost less and are more effective.

Microwave Detectors

Microwave detectors work like ultrasonic detectors, but they send high-frequency radio waves instead of sound waves. Unlike ultrasonic waves, these microwaves can go through walls and be shaped to protect areas of various configurations. Microwave detectors are easy to conceal because they can be placed behind solid objects. When properly adjusted, they are not susceptible to loud noises or air movement.

The big drawback to microwave detectors is their sensitivity makes them hard to adjust properly. Because the waves penetrate walls, a passing car can prompt a false alarm. Their alarms can also be triggered by fluorescent lights or radio transmissions. Microwave detectors are rarely useful for homes.

Passive Infrared Detectors

Passive infrared (PIR) detectors became popular in the 1980s. Today, they are the most cost-effective type of interior device for homes. A *PIR* senses rapid changes in temperature within a protected area by monitoring infrared radiation (energy in the form of heat). A PIR uses less power, is smaller, and is more reliable than either an ultrasonic or a microwave detector.

The PIR is effective because all living things give off infrared energy. If an intruder enters a protected area, the device senses a rapid change in heat. When properly adjusted, the detector ignores all gradual fluctuations of temperature caused by sunlight, heating systems, and air conditioners.

A typical PIR can monitor an area measuring about 20 by 30 feet, or a narrow hallway about 50 feet long. It doesn't penetrate walls or other objects, so a PIR is easier to adjust than a microwave detector and it doesn't respond to radio waves, sharp sounds, or sudden vibrations.

The biggest drawback to PIRs is they don't "see" an entire room. They have detection patterns made up of "fingers of protection." The spaces outside and between the "fingers" aren't protected by the PIR. How much of an area is monitored depends on the number, length, and direction of zones created by a PIR's lens and on how the device is positioned.

Many models have interchangeable lenses that offer a wide range of detection pattern choices. Some patterns, called "pet alleys," are several feet above the floor to allow pets to move about freely without triggering the alarm. Which detection pattern is best for you depends on where and how your PIR is being used.

A useful feature of the latest PIRs is *signal processing* (also called *event verification*), which can reduce false alarms by distinguishing between large and small differences in infrared energy.

Quads

A quad PIR (or *quad*, for short) consists of two dual-element sensors in one housing. Each sensor has its own processing circuitry, so the device is basically two PIRs in one. A quad reduces false alarms because, to trigger an alarm, both PIRs must simultaneously detect an intrusion. That feature prevents the alarm from activating in response to insects or mice. A mouse, for example, may be detected by the fingers of protection of one of the PIRs, but it would be too small to be detected by both at the same time.

Dual Techs

Detection devices that incorporate two different types of sensor technology into one housing are called dual technology devices (or dual techs). A *dual tech* triggers an alarm only when both technologies sense an intrusion. Dual techs are available for commercial and residential use, but because they can cost several hundred dollars each, dual techs are more often used by businesses. The most effective dual tech for most homes is one that combines PIR and microwave technology.

For this type of dual tech to trigger an alarm, a condition must exist that simultaneously triggers both technologies. The presence of infrared energy alone would not trigger an alarm. Movement outside a wall, which might ordinarily trigger a microwave, for example, won't trigger a dual tech because the PIR element wouldn't simultaneously sense infrared energy.

Types of Alarms

There are two basic types of intruder alarms: wireless and hardwired. Both consist of a control panel and a *keypad* or *keyswitch*, which is a sounding device (such as a bell or siren). The big difference is, with a *wireless system*, you don't use wire to connect the sensors to the control panel. The control panel is basically a radio receiver and the sensors are transmitters.

Although a wireless system is generally easier to install, it's more prone to false alarms than a hardwired system. And, when a wireless system has a false alarm, you may have a hard time determining the cause. Lightning, airplanes, and electronic garage door openers can cause false alarms. With a *hardwired system*, a false alarm can often be traced to a broken or loose wire.

With either type of system, the control panel is the brain: it holds programmed information that tells the system how to function. The control panel is usually the most expensive part of an alarm system.

The sensors are strategically positioned throughout a home to detect an intrusion and to alert the control panel. The system can be armed and disarmed with either a keypad or a keyswitch.

In some small systems, the keypad and control panel are one unit. Generally all the parts of an alarm system can be bought separately, and then mixed and matched with components from other manufacturers. You can buy a control panel from one manufacturer and sensors from others.

More About Sensors

Sensors (or *detection devices*) are installed on doors, windows, and floors to detect sound, air movement, or body heat. When a person enters a protected area, the sensors inform the control panel. Depending on how the panel is set to react, it may wait from 15 seconds to a minute before activating the system's sirens and lights, and then calling for help. A delay is important so that when an authorized person enters the home, they have time to disarm the system.

Dozens of different sensors are on the market. They cost from a few cents to hundreds of dollars. Buying them based on price alone, however, can be penny wise and pound foolish. Each type of sensor works best under certain conditions. To keep false alarms to a minimum, you need to use the right sensors.

Sensors come in two basic types. A *perimeter sensor* is installed on a door, window, or other opening to detect intruders before they enter a protected area. An *interior* (or *space*) *sensor* protects such open areas as rooms and hallways. It's used to detect an intruder who gets past perimeter sensors.

The most popular perimeter sensor is the magnetic switch. As its name implies, the *magnetic switch* consists of a magnet and a switch, housed in a matching case. You can buy magnetic switches for a few dollars per set.

In a typical installation, the magnet is mounted on the edge of a door or window, and the switch is aligned about 1/2-inch away on the frame. If someone pushes the door or window open, the magnet moves out of alignment, activating the alarm.

Three popular types of interior sensors are microwaves, passive infrareds, and dual technology. A microwave detector emits high-frequency radio waves. When an intruder enters a protected area, the device senses a change in radio-wave pattern that triggers an alarm. Because radio waves can penetrate walls (and other solid objects), the detector can easily be hidden in a room.

However, the microwave's detection sensitivity is hard to adjust accurately, making it highly susceptible to false alarms. A microwave detector has been known to mistake many things for intruders, including passing cars, fluorescent lights, and radio transmissions.

One of the latest types of sensors is dual technology (or dual tech), which combines two sensing technologies into one housing and triggers an alarm only when both simultaneously sense an intrusion. This reduces false alarms.

The most popular type of dual tech combines a PIR with a microwave detector. See Figure 11.2. Passing cars and other outside movement that might affect the microwave won't affect the PIR. Conversely, a heater or sunlight on a window might affect the PIR, but they won't trigger the microwave. An intruder who walks into a protected area, however, can trigger both technologies.

The problem with dual techs is price. Some models cost several hundred dollars (two or three times more than PIRs or microwaves alone), but using only the most expensive sensors is seldom necessary. Professional installers strategically combine several types.

Monitoring an Alarm

An alarm can be monitored locally or remotely. *Local monitoring* means the homeowner/renter or a neighbor listens for the

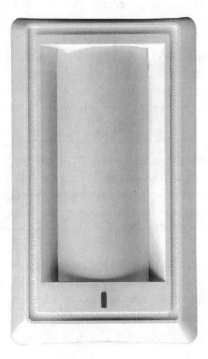

Figure 11.2   A passive infrared (PIR) is one of the most common sensors used in intruder alarms (courtesy of Napco).

alarm. *Remote monitoring* means someone at another location is notified when the alarm is activated.

The least-costly way to have remote monitoring is to connect a tape dialer to the system. When the alarm is activated, the dialer uses the telephone to call one or more preprogrammed numbers and delivers a recorded message. The idea is that someone will get the message and call the police. (In some cities, the police respond directly to the recorded message.)

The big problem with a tape dialer is there's no way to know whether the system was activated by a false alarm. Too many false alarms upset friends, neighbors, and the police. Some police departments levy fines for repeat false alarms.

The best way to avoid repeatedly disturbing friends and neighbors is to use the services of a central monitoring station. When an alarm is activated, a digital dialer sends a coded electronic signal over the phone line to the monitoring station, which is manned 24 hours a day.

Depending on the signal received, the monitoring station either immediately calls the police or fire department, or calls the homeowner/renter to verify the emergency. If the monitoring station calls the person and no one answers, or if someone answers without giving the unique code word, the operator politely hangs up and calls the police.

Such an arrangement enables the homeowner/renter to inform the caller in the case of a false alarm, but it prevents an intruder from stopping the monitoring station from calling for help.

If you install an alarm, try to convince the person to use a monitoring station. Because you can bill monthly for the service, this is a good way of getting recurring revenue. For a list of monitoring stations, see Appendix E.

Installing Wireless and Hardwired Systems

Be sure to read and understand the manufacturer's instructions before you start to install a system. If you don't understand something, you can usually call a technical support line or get technical information from the manufacturer's web site. The tips in this chapter are general guidelines and aren't meant to take the place of the manufacturer's instructions.

One of the first considerations in choosing components for an alarm system is cost. You must decide how to provide adequate protection while making a good profit. Generally, that doesn't include using a system of redundant state-of-the-art perimeter sensors at every entry point, backed up by overlapping high-technology interior sensors that saturate the premises. Using the best of everything is usually overkill and more than customers want to pay for. Under no circumstances, however, should you install a system that you know provides inadequate protection. Regardless of what you charge, you should always give a reasonable minimum standard of quality. If customers can't afford that, then let them go elsewhere. Otherwise, you're likely to face lawsuits and you could get a bad reputation in your community.

To plan an effective alarm system, you need to become familiar with the layout of the premises or areas to be protected, as well as the needs and limitations of the people who are going to use the system. Talk with members of the household or business where the system is being installed. Don't make the mistake of doing cookie-cutter, one-size-fits-all types of installations. To compete with the giants in the alarm industry, you can't copy them. You have to offer superior personalized service.

Determine what your customers expect from a system. Ask about their concerns, and find out who will use the system. Some people may have special needs, such as young children or persons with a disability. The more you learn about them, the greater chance you can educate them. An educated customer is a good customer. When customers keep causing false alarms or get frustrated while trying to figure out how to use the system, they won't blame themselves. They'll blame you.

Once you understand what the customer wants and can likely work with, you need to draw a sketch of the areas to be protected. The sketch should show all doors, windows, and other openings, along with all rooms. Indicate at each area to be protected which components and sensors you plan to use. Also, decide how many zones you want, and which components will be in which zones. When making the drawing, keep in mind how you'll run any wire that may be needed. The better you plan, the easier your installation.

Installing a Wireless Alarm

Many wireless alarms combine the control panel with the keypad for easy wall mounting or placement on a table or desk. If the control panel is wall-mounted, make sure it's at a comfortable height for your customers. The control panel should be installed inside a building near the most-often-used entrance. It shouldn't be visible from outside through a window or door.

Install the interior bell or siren in an inconspicuous place, where it can be heard throughout the house or throughout much of the building. Don't place it where the sound will be muffled by draperies or furniture. An X-10 compatible (the most popular type of wireless alarm) siren plugs into an electrical outlet and communicates with the other components over the building's wiring (don't plug this alarm into an outlet controlled by a switch). Then, use a small screwdriver to turn the dials to choose a *house code*, to which you program all the other X-10 compatible components.

Next, program the control panel. Some systems use a synthesized voice to talk you through the programming steps. Then, install all the sensors and test the system. Have members of the household or business work with the system, and make sure they know how to use it.

Installing Hardwired Alarms

To install a hardwired system, you need to run wire from the control panel to various components. This is why hardwired systems are far less popular among do-it-yourselfers. But running wire doesn't have to be hard. In general, hardwired systems are more reliable and easier to troubleshoot than a wireless system. See Figure 11.3.

What type and size wire to use, and how to run it, are determined by requirements of the National Electrical Code (NEC), the component manufacturer, your local codes, the importance of aesthetics, the amount your customer is willing to pay, and the structure of the building. The wiring for most hardwired systems is from 18 gauge to 22 gauge. You want to get the wire to places where it can't easily be seen. To hide wiring, it needs to be run from one floor to others and within a room, while exposing as little wire as possible. That's for aesthetic as well as security reasons.

Wiring can be run open or concealed (or a combination of both). In *open wiring*, the wire is run where it's in plain view and easily accessible, such as on the surface of walls, columns,

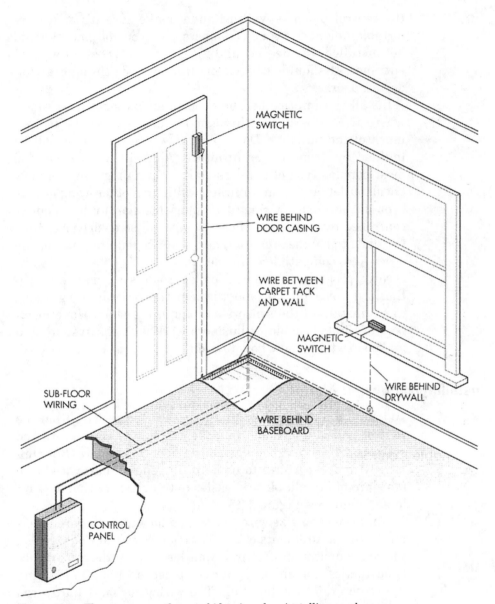

Figure 11.3   There are many places to hide wire when installing an alarm.

and ceiling. If you run open wiring, run it through conduit. That kind of wiring doesn't look very neat.

In *concealed wiring*, the wire is run within walls, behind baseboards, over drop ceilings, inside of columns, and otherwise out of view. Concealed wiring is the more popular type of wiring for homes and other places where appearance is important. When you're running wire behind moldings or casings, consider cutting a cavity beneath them where the wire can be run. If the molding or baseboard is painted, before prying it loose, use a utility knife to cut between the wall and the molding or baseboard. That helps keep the paint from peeling. Then, use a flat pry bar to carefully work it off. If the nails in the molding or baseboard aren't too bent, leave them in place, so you can align them with their original holes later. This makes replacement easier.

Try not to run wire across finished walls through studs, because it requires a lot of cutting and patching. When you are considering where to run wire, look for existing holes and spaces. Most buildings have many you can use. Good places to run wire include an unfinished attic, a crawlspace, a basement, closets, and over drop ceilings. You may be able to run wire along the plumbing stack from the attic to the basement, by lowering a fishing weight attached to a thin chain or string. When it reaches the basement, have a helper attach a cable to the chain or string. Pull the cable back past the stack, and you then have a cable from the basement to the attic.

If you have to drill holes, look for inconspicuous places, such as a closet that's directly above another closet. Other options include stapling wire along the edges of carpet and hiding it behind baseboards, window casings, and moldings.

When you can't hide wire, you may need to use a *raceway*, which is a channel made of metal or insulating material used for holding wire and cable. Some raceways include electrical metal tubing (EMT), rigid metal conduit, rigid nonmetallic conduit, intermediate metal conduit (IMC), and flexible metal conduit.

Corrosion-resistant metal conduit and fittings are used when corrosion might be a problem. Plastic-coated steel, aluminum, and silicon bronze alloy are corrosion-resistant. If the location where the conduit is being installed is wet, space the conduit at least 1/4-inch away from the mounting surface.

Lengths of conduit are installed with a threaded coupling at one end. To cut conduit to size, you can use a hacksaw, a power band saw, or a conduit cutter. After cutting, you need to ream the conduit to remove any burrs and sharp edges. Then, lubricate the conduit with cutting oil and run a cutting die up the conduit until you have a full thread.

12

Electromagnetic Locks

The electromagnetic lock was introduced in the United States in 1970 and has gained considerable popularity. Today, it's a popular part of access control systems throughout the world.

Electromagnetic locks are often used to secure emergency exit doors. When connected to a fire alarm system, the lock's power source is automatically disconnected when the fire alarm is activated. That allows the door to open freely, so people can exit quickly.

Although the principle of operation of an electromagnetic lock is different from that of a conventional mechanical lock, the former has proven a cost-effective, high-security locking device. Unlike a mechanical lock, an electromagnetic lock doesn't rely on the release of a bolt or a latch for security. Instead, it relies on electricity and magnetism.

A standard electromagnetic lock consists of two components: a rectangular electromagnet and a rectangular, ferrous metal strike plate. See Figure 12.1. The electromagnet is installed on a door's header; the strike plate is installed on the door in a position that allows it to meet the electromagnet when the door closes. When the door is closed and the electromagnet is adequately powered (usually by 12 to 24 dc volts at 3 to 8 watts), the door is secured. Typically, the locks have 300 to 3,000 pounds of holding power.

One of the biggest fears people have about electromagnetic locks is power failure. What happens if the power goes out or

Figure 12.1 A standard electromagnetic lock (courtesy of Highpower Security Products, LLC).

if a burglar cuts the wire connecting the power source to the lock? Standby batteries are often installed with the lock to provide continued power in such cases. Also, the lock can't be tampered with from outside the door because it is installed entirely inside the door. No part of the lock or power supply wires is exposed from outside the door.

Another important security feature of electromagnetic locks is they are fail-safe. That is, when no power is going to the electromagnet, the door will not be locked. That is why the lock meets the safety requirements of many North American building codes.

Electromagnetic locks have two major disadvantages. First, the locks often cost from four to ten times more than typical high-security mechanical locks. And, second, many people think electromagnetic locks are much less attractive than mechanical locks.

When using electromagnetic locks in a typical application, additional pieces of hardware must be installed to comply with building fire codes. These fire codes are used throughout the United States. The code states the following: There must be a minimum of two devices used to release the electromagnetic lock. One device must be a manual release button that has the words "PUSH TO EXIT" labeled. This push button must provide a 30-second time delay when pushed, and the time delay must act independently of the access control system (the delay must work on its own, not tied into any other access control system). Another device can either be a PIR motion detector (as shown in Figure 12.2) or an electrified exit release bar (also called a "crash bar"). If the building has a fire alarm system, the electromagnetic lock must be tied into the fire control sys-

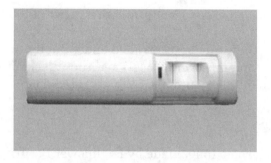

Figure 12.2   A request to exit PIR motion sensor (courtesy of Highpower Security Products, LLC).

tem, so the lock unlocks automatically during a fire alarm. The door must also release upon the loss of main electrical power.

The DS-1200 Electromagnetic Lock

The DS-1200 model, made by Highpower Security Products, has 1,200 pounds of holding power, and is for external or industrial applications. It can be used in harsh environments to secure doors and gates. All electronics are sealed in epoxy and are protected by the steel housing cover. The housing armature and exposed face of the electric lock are nickel plated to resist rust and corrosion. A rigid conduit fitting can be provided on one end of the lock to protect power wiring in gate control installations.

All DS-1200 electromagnetic locks are fail-safe, releasing instantly on command or loss of power. There are no moving parts to wear, stick, or bind.

The rugged design and durable construction of this lock assures virtually unlimited actuations without fear of electrical fatigue or mechanical breakdown. Proudly made in the U.S.A., a ten-year limited warranty is provided.

The standard DS-1200 model is supplied with an adjustable mounting plate for use on out-swinging doors. The DS-1200-TJ unit is furnished with an angle lock mounting plate and an armature Z bracket for in-swinging door installations. Any 1200 series lock may be converted in the field for in-swinging door applications by adding the angle mounting plate and Z bracket. An optional conduit fitted lock is available for exterior gate control applications.

All 1200 series locks can be operate on either 12 or 24 VDC. The efficient design of these locks requires only 170 milliamp (ma) at 24 volts dc to maintain the rated 1,200-lb. holding force.

Because they can be controlled individually or simultaneously from one or several locations, these locks are ideal for securing manual or automatically operated doors and gates.

The Thunderbolt 1500

The Thunderbolt 1500, made by Highpower Security Products, is a 2-inch profile electromagnetic lock for out-swinging doors that combines unmatched ease of installation with solid product reliability. Designed to provide the fastest installation, the unit incorporates a unique tamper-resistant cover design and slotted mounting system with installation template. Providing 1,500 lbs. of holding force, all versions feature a replaceable dual-voltage coil that operates at both 12 and 24 volts. Units can be equipped to operate using either ac or dc power, and they feature a surge- and spike-suppression circuit. The Thunderbolt 1500 is made in the U.S. and is backed with a ten-year manufacturer's warranty.

The lock is available in over 150 colors of powder-coat finishes and in solid brass by special order. This is a slim-line 1,500-lb. electromagnetic lock, designed to provide the fastest installation times. Mounting the Thunderbolt is quick and easy! Extensive feedback from installers directed Highpower to incorporate an improved adjustable mounting system. This slotted mounting system allows freedom of movement to make adjustments during tough installations. A template is provided to quickly mark mounting holes for both the magnet and the armature.

Installers are discovering that the Thunderbolt 1500 is designed to maximize profits. The Thunderbolt 1500 has an epoxy-less design that allows the magnetic coil to be unplugged from the unit and replaced if it becomes defective, without having to uninstall the lock. This feature both reduces service time and provides improved value, by keeping assembly costs low.

No more fooling around with wire splices! Connections are made to the Thunderbolt with screw terminal blocks that speed wire installation. With a single circuit board, the Thunderbolt can quickly be configured with a door position switch (DPS), a cover tamper switch (CTS), and a magnetic bond sensor (MBS). In addition, electronic spike and "kick-back" surge suppression is standard with all models.

Installers love the Thunderbolt's single-piece cover. It slides into place, allowing a rapid and hassle-free installation. Because it has no exposed screws, the cover provides the highest level of tamper resistance. This modular cover allows installers to stock different color covers to quickly provide customers with the cosmetics specified.

M32 600lb Holding Force Magnalock

Application - 600 lb. holding force Magnalock recommended for applications where physical assault on the door is not expected, like access controlled interior rooms and secure areas within buildings. Specify 12 or 24 VDC when ordering.

Features and Benefits

- Instant release circuit - no residual magnetism
- Surface mounts easily with a minimum of tools
- Fully sealed electronics - tamper proof and weatherproof
- Mounted using steel machine screws into blind finishing nuts
- Architectural brushed stainless steel finish (US32D/630)
- All ferrous metal surfaces plated to MIL specification
- Hardware accessories available to configure any opening type include mounting brackets, housings and dress covers
- Ten feet [3.05m] of jacketed, stranded conductor
- Also available in Senstat and Face Mount models
- **UL Listed**

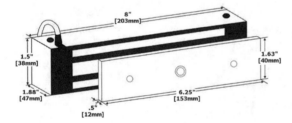

Specification Data

Holding Force: 600 lbs. [272 Kg.]
Current Draw and Voltage:
300mA at 12VDC; 150mA at 24VDC
Operating Temperature:
-40 to +140F [-40 to +60C]
Warranty: MagnaCare Lifetime Replacement Warranty

How To Order - 12VDC

Part #	Description
M32-12	Magnalock Model 32 12VDC
M32SC-12	Model 32 Senstat 12VDC
M32F-12	Model 32 Face Mount 12VDC
M32SCF-12	Model 32 Senstat Face Mount 12VDC

How To Order - 24VDC

Part #	Description
M32-24	Magnalock Model 32 24VDC
M32SC-24	Model 32 Senstat 24VDC
M32F-24	Model 32 Face Mount 24VDC
M32SCF-24	Model 32 Senstat Face Mount 24VDC

M34 600lb Holding Force Magnalock

Application - 600 lb. holding force Magnalock with automatic dual voltage. Recommended for applications where physical assault on the door is not expected, like access controlled interior rooms and secure areas within buildings.

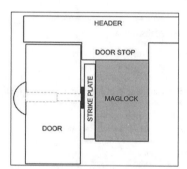

Features and Benefits

- Instant release circuit - no residual magnetism
- Surface mounts easily with a minimum of tools
- Fully sealed electronics - tamper proof and weatherproof
- Mounted using steel machine screws into blind finishing nuts
- Architectural brushed stainless steel finish (US32D/630)
- All ferrous metal surfaces plated to MIL specification
- Hardware accessories available to configure any opening type include mounting brackets, housings and dress covers
- Ten feet [3.05m] of jacketed, stranded conductor
- Automatic dual voltage - no field adjustment required
- Also available in Senstat and Face Mount models
- **UL Listed**

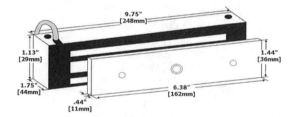

Specification Data

Holding Force: 600 lbs. [272 Kg.]
Current Draw and Voltage:
350mA at 12VDC; 175mA at 24VDC
Operating Temperature:
-40 to +140F [-40 to +60C]
Warranty: MagnaCare Lifetime Replacement Warranty

How To Order

Part #	Description
M34	Magnalock Model 34
M34SC	Model 34 Senstat
M34F	Model 34 Face Mount
M34SCF	Model 34 Senstat Face Mount

M34R 600lb Holding Force Magnalock

Application - 600 lb. holding force Magnalock with automatic dual voltage for mounting in sliding doors. This results in an aesthetically appealing finished appearance. Recommended for applications where physical assault on the door is not expected, like access controlled interior rooms and secure areas within buildings.

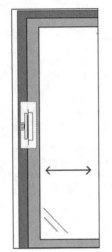

Features and Benefits

- Instant release circuit - no residual magnetism
- Recessed for aesthetically appealing finish
- Fully sealed electronics - tamper proof and weatherproof
- Mounted using steel machine screws into blind finishing nuts
- Architectural brushed stainless steel finish (US32D/630)
- All ferrous metal surfaces plated to MIL specification
- Ten feet [3.05m] of jacketed, stranded conductor
- Automatic dual voltage - no field adjustment required
- Also available in Senstat model
- **UL Listed**

Specification Data

Holding Force: 600 lbs. [272 Kg.]
Current Draw and Voltage:
350mA at 12VDC; 175mA at 24VDC
Operating Temperature:
-40 to +140F [-40 to +60C]
Warranty: MagnaCare Lifetime
Replacement Warranty

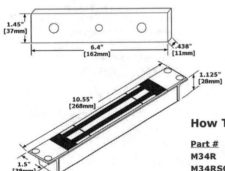

How To Order

Part #	Description
M34R	Magnalock Model 34 - Recessed
M34RSC	Magnalock Model 34 - Recessed Senstat

M38 600lb Holding Force Magnalock

Application - 600 lb. holding force Magnalock with automatic dual voltage. This model offers a built in wire access chamber and new bracket mount design for a unique method of installation. Recommended for applications where physical assault on the door is not expected, like access controlled interior rooms and secure areas within buildings.

Features and Benefits

- Instant release circuit - no residual magnetism
- Surface mounts easily with mounting bracket using a minimum of tools
- Internal wire access chamber
- Mounted using steel machine screws into blind finishing nuts
- Brushed aluminum finish
- All ferrous metal surfaces plated to MIL specification
- Automatic dual voltage - no field adjustment required
- Dress covers available
- **UL Listed**

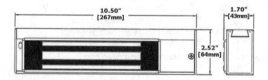

Specification Data

Holding Force: 600 lbs. [272 Kg.]
Current Draw and Voltage:
350mA at 12VDC; 175mA at 24VDC
Operating Temperature:
-40 to +140F [-40 to +60C]
Warranty: MagnaCare Lifetime Replacement Warranty

How To Order

Part #	Description
M38	Magnalock Model 38
M38D	Model 38 Door Position
M38DL	Model 38 Door Position, LED
M38DLS	Model 38 Door Position, LED, Senstat
M38DLST	Model 38 Door Position, LED, Senstat, Tamper
M38DLT	Model 38 Door Position, Tamper
M38DS	Model 38 Door Position, Senstat
M38DST	Model 38 Door Position, Senstat, Tamper

How To Order

Part #	Description
M38DT	Model 38 Door Position, Tamper
M38L	Model 38 LED
M38LS	Model 38 LED, Senstat
M38LT	Model 38 LED, Tamper
M38S	Model 38 Senstat
M38ST	Model 38 Senstat, Tamper
M38T	Model 38 Tamper

See page 19 for a full explanation of the M38 Options

M62 1200lb Holding Force Magnalock

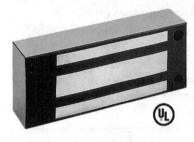

Application - 1200 lb. holding force Magnalock with automatic dual voltage. Recommended for exterior and perimeter doors through which criminals could gain entrance to cause substantial economic loss.

Features and Benefits

- Instant release circuit - no residual magnetism
- Fully sealed electronics - tamper proof and weatherproof
- Mounted using steel machine screws into blind finishing nuts
- Architectural brushed stainless steel finish (US32D/630)
- All ferrous metal surfaces plated to MIL specification
- Hardware accessories available to configure any opening type include mounting brackets, housings and dress covers
- Ten feet [3.05m] of jacketed, stranded conductor
- Automatic dual voltage - no field adjustment required
- Also available in Senstat, Conduit, Double and Face Mount models
- **UL Listed**

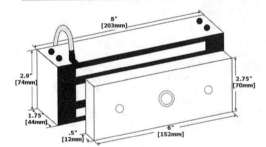

Specification Data

Holding Force: 1,200 lbs. [544 Kg.]
Current Draw and Voltage:
250mA at 12VDC; 125mA at 24VDC
Operating Temperature:
-40 to +140F [-40 to +60C]
Warranty: MagnaCare Lifetime Replacement Warranty

How To Order

Part #	Description
M62	Model 62 Magnalock
M62F	Model 62 Face Mount Magnalock
M62FG	Model 62 Face Mount Conduit Mag.
M62G	Model 62 Conduit Magnalock
M62G-OS	Model 62 Conduit Offset Strike Mag.
M62G-SS	Model 62 Conduit Split Strike Mag.
M62-OS	Model 62 Offset Strike Magnalock
M62SC	Model 62 Senstat Magnalock

How To Order

Part #	Description
M62SCF	Model 62 Senstat Face Mount Magnalock
M62SCFG	Model 62 Senstat Face Mount Conduit Mag.
M62SCG	Model 62 Senstat Conduit Magnalock
M62SCG-OS	Model 62 Senstat Conduit Offset Strike Mag.
M62SCG-SS	Model 62 Senstat Conduit Split Strike Mag.
M62SC-OS	Model 62 Senstat Offset Strike Magnalock
M62SC-SS	Model 62 Senstat Split Strike Magnalock
M62-SS	Model 62 Split Strike Magnalock

DM62 Double Magnalock

Application - For double doors without a center mullion, the DM62 supplies two model 62 Magnalocks in a single, brushed stainless steel housing.

Features and Benefits

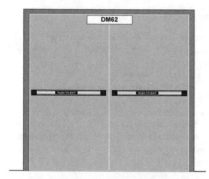

- Instant release circuit- no residual magnetism
- Electrically, both locks can be controlled as one or separately
- Surface mounts easily with a minimum of tools
- Fully sealed electronics - tamper proof and weather proof
- Mounted using steel machine screws into blind finishing nuts
- Architectural brushed stainless steel finish (US32D/630)
- All ferrous metal surfaces plated to MIL specification
- Hardware accessories available to configure any opening type include mounting brackets
- Ten feet [3.05m] of jacketed, stranded conductor for each lock
- Automatic dual voltage - no field adjustment required
- Also available in Senstat model
- **UL Listed**

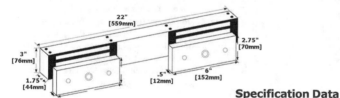

Specification Data

Holding Force: 1,200 lbs. [544 Kg.] per lock
Current Draw and Voltage:
250mA at 12VDC; 125mA at 24VDC per lock
Operating Temperature:
-40 to +140F [-40 to +60C]
Warranty: MagnaCare Lifetime Replacement Warranty

How To Order

Part #	Description
DM62	M62 Double Magnalock
DM62SC	M62 Double Magnalock Senstat

M68 1200lb Holding Force Magnalock

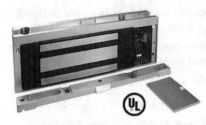

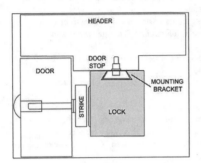

Application - 1200 lb. holding force Magnalock with automatic dual voltage. This model offers a built in wire access chamber and new bracket mount design for a unique method of installation. Recommended for exterior and perimeter doors through which criminals could gain entrance to cause substantial economic loss.

Features and Benefits

- Instant release circuit - no residual magnetism
- Surface mounts easily with mounting bracket using a minimum of tools
- Internal wire access chamber
- Mounted using steel machine screws into blind finishing nuts
- Brushed aluminum finish
- All ferrous metal surfaces plated to MIL specification
- Automatic dual voltage - no field adjustment required
- Dress covers available
- **UL Listed**

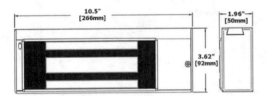

Specification Data

Holding Force: 1200 lbs. [544 Kg.]
Current Draw and Voltage:
300-325mA (options) at 12VDC
150-175mA (options) at 24VDC
Operating Temperature:
-40 to +140F [-40 to +60C]
Warranty: MagnaCare Lifetime
Replacement Warranty

How To Order

Part #	Description
M68	M68 Magnalock
M68D	M68 Door Position
M68DL	M68 Door Position, LED
M68DLS	M68 Door Position, LED, Senstat
M68DLST	M68 Door Position, LED, Senstat, Tamper
M68DLT	M68 Door Position, LED, Tamper
M68DS	M68 Door Position, Senstat
M68DST	M68 Door Position, Senstat, Tamper

How To Order

Part #	Description
M68DT	M68 Door Position, Tamper
M68L	M68 LED
M68LS	M68 LED, Senstat
M68LT	M68 LED, Tamper
M68S	M68 Senstat
M68ST	M68 Senstat, Tamper
M68T	M68 Tamper

Additional M68 part numbers listed on the following page

M82 1800lb Holding Force Magnalock

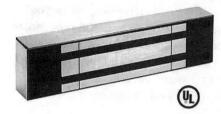

Application - 1800 lb. holding force Magnalock with automatic dual voltage and Senstat monitoring function standard. Preferred by prisons and detention facilities, and also functions best on large, high security gates.

Features and Benefits

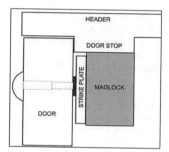

- Instant release circuit - no residual magnetism
- Surface mounts easily with a minimum of tools
- Fully sealed electronics - tamper proof and weatherproof
- Mounted using steel machine screws into blind finishing nuts
- Architectural brushed stainless steel finish (US32D/630)
- All ferrous metal surfaces plated to MIL specification
- Hardware accessories available to configure any opening type include mounting brackets, housings and dress covers
- Ten feet [3.05m] of jacketed, stranded conductor
- Automatic dual voltage - no field adjustment required
- Also available in Conduit and Face Mount models
- **UL Listed**

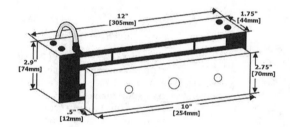

Specification Data

Holding Force: 1,800 lbs. [816 Kg.]
Current Draw and Voltage:
350mA at 12VDC; 175mA at 24VDC
Operating Temperature:
-40 to +140F [-40 to +60C]
Warranty: MagnaCare Lifetime Replacement Warranty

How To Order

Part #	Description
M82SC	Magnalock Model 82 Senstat
M82SCF	M82 Senstat Face Mount
M82SCFG	M82 Senstat Face Mount Conduit
M82SCG	M82 Senstat Conduit

How To Order

Part #	Description
M82SCG-OS	M82 Senstat Conduit Offset Strike
M82SCG-SS	M82 Senstat Conduit Split Strike
M82SC-OS	M82 Senstat Offset Strike
M82SC-SS	M82 Senstat Split Strike

SAM-Shear Aligning Magnalock

Application - 1200 lb. holding force Shear Aligning Magnalock mounts fully concealed into wood, steel or aluminum doors. SAM can also be installed in any position on the door - top, side or bottom and is ideal for swing through and sliding doors.

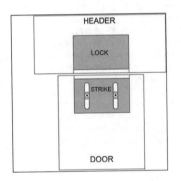

Features and Benefits

- Instant release circuit - no residual magnetism
- Mounts easily with a minimum of tools
- Fully sealed electronics - tamper proof and weatherproof
- Mounted using steel machine screws
- Architectural brushed stainless steel finish (US32D/630)
- All ferrous metal surfaces plated to MIL specification
- Hardware accessories available to configure any opening type include mounting brackets
- Ten feet [3.05m] of jacketed, stranded conductor
- Automatic dual voltage - no field adjustment required
- **UL Listed**

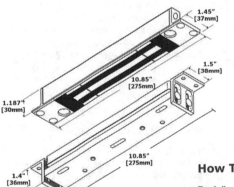

Specification Data

Holding Force: 1,200 lbs. [544 Kg.]
Current Draw and Voltage:
350mA at 12VDC; 175mA at 24VDC
Operating Temperature:
-40 to +140F [-40 to +60C]
Warranty: MagnaCare Lifetime Replacement Warranty

How To Order

Part #	Description
SAM	Shear Aligning Magnalock
SAMSC	Shear Aligning Magnalock Senstat

SAM2-24- Shear Aligning Magnalock

Application - 600 lb. holding force Shear Aligning Magnalock mounts fully concealed and is ideal for smaller applications, like cabinet doors and sliding closet doors. 24 VDC only.

Features and Benefits

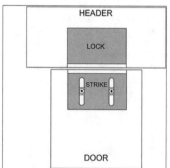

- Instant release circuit - no residual magnetism
- Mounts easily with a minimum of tools
- Fully sealed electronics - tamper proof and weatherproof
- Mounted using steel machine screws
- Architectural brushed stainless steel finish (US32D/630)
- All ferrous metal surfaces plated to MIL specification
- Ten feet [3.05m] of jacketed, stranded conductor
- Smaller dimensions for difficult installation areas
- **UL Listed**

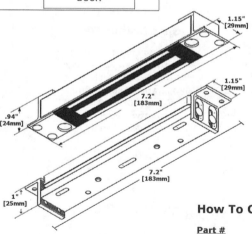

Specification Data

Holding Force: 600 lbs. [272 Kg.]
Current Draw and Voltage: 62mA at 24VDC only
Operating Temperature: -40 to +140F [-40 to +60C]
Warranty: MagnaCare Lifetime Replacement Warranty

How To Order

Part #	Description
SAM2-24	SAM2 Magnalock 24VDC

MCL 200lb Holding Force Magnalock

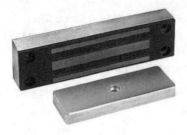

Application - 200 lb. holding force. Typical applications include: jewelry cases, museum cases, gun cabinets, pharmaceutical storage cabinets, cash drawers and any sort of door or drawer which will physically accommodate the lock.

Features and Benefits

- Instant release circuit - no residual magnetism
- Surface mounts easily with a minimum of tools
- Fully sealed electronics - tamper proof and weatherproof
- Mounted using steel machine screws
- Architectural brushed stainless steel finish (US32D/630)
- All ferrous metal surfaces plated to MIL specification
- Six feet [1.8m] of jacketed, stranded conductor

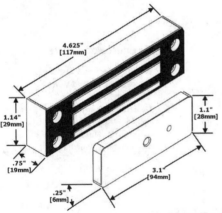

Specification Data

Holding Force: 200 lbs. [91 Kg.]
Current Draw and Voltage:
62mA at 24VDC only
Operating Temperature:
-40 to +140F [-40 to +60C]
Warranty: MagnaCare Lifetime
Replacement Warranty

How To Order

Part #	Description
MCL-24	Magnetic Cabinet Lock 24VDC

13

Closed-Circuit Television Systems

This chapter looks at the various components that make up a closed-circuit television (CCTV) system, and it shows various ways that such a system can be used.

Locks and other physical security devices can be more effective when used in conjunction with a CCTV system. Such a system can allow numerous areas—elevators, entrances and exits, parking lots, lobbies, and cash-handling areas—to be monitored constantly. Such monitoring can deter crime and reduce a company's security costs.

Basics

To begin selling and installing CCTV systems, you don't need to have a strong background in electronics. Although few locksmiths can handle the large multiplex systems used in airports and banks, most CCTVs used in small stores, offices, and apartments are easy to install. Many CCTVs come preconfigured as a complete package.

However, it's still important for you, the security professional, to have a basic understanding of each of the components you may be working with. That enables you to be more helpful to your customer—which can mean more profits for you. The major parts of a CCTV system are the cameras, housings, monitors, video recorders, and pan-and-tilt drives.

Cameras

There are two main types of CCTV cameras: vacuum tube (tube) and solid state CCD (chip). Tube cameras are inexpensive, but they have a short lifespan. CCD cameras not only last longer, they also provide better resolution and adapt better to varying light conditions. As a rule, CCD cameras are a much better value than their tube counterparts.

Cameras come in color or black-and-white. Black-and-white models are less expensive, and they are better for outdoor and low-light applications.

Cameras can be overt or covert. *Overt models* are designed to let people know they may be being watched. Some are stylishly designed to fit various décor. *Covert models* are designed to be unnoticed. Some covert models have been innocuously build into common items, such as wall clocks, radios, smoke detectors, and exit signs.

Housings

Special housings are available to protect cameras from vandalism, weather, and environmental conditions, such as dust. Accessories incorporated into the housings include batteries, hoods, and lens-cover wipers. Housings are generally made of metal or high-impact plastic.

Monitors

Like cameras, monitors are available in color and black-and-white. Unless you use a color camera, there's no need to use a color monitor. (To view a scene in color, both the camera and the monitor must be color models.) Some monitors include a built-in switcher, which lets you automatically or manually view scenes from multiple cameras. Monitors range in size from 5 inches to over 23 inches.

Video Recorders

A CCTV system has limited value if it can't record select events. Two basic types of video recorders are video tape and video cassette. Reel-to-reel video tape recorders can provide both real-time and time-lapse recording. However, video cassette recorders are easier to use and they are popular for security applications. Standard video tape recorders can record continuously for up to six hours. Time-lapse models can record up to 200 hours. Tape cost can be kept low by periodically erasing and reusing tapes.

Pan-and-Tilt Drives

A *pan-and-tilt drive* is a motor-driven device that holds a camera. The device lets you remotely move the camera to various positions.

Installing a CCTV System

To install a CCTV system, you need to run cable from the cameras to the monitors. Typical cable includes fiber-optic, coaxial, and twisted pair. If you don't know how to run cable, see the section on running wire in Chapter 11.

14

Access Control Systems

As the name implies, an *access control system* controls the flow of pedestrian or vehicular traffic through entrances and exits of a protected area or premises. Basically, it's an electronic way of controlling who can go where and when. Such a system typically uses coded cards, biometric readers, or magnetic keys.

Card systems are often used at hotels for entering rooms. The coded information on the key card works only to the extent that the person controlling the access control system wants it to. They can make the card invalid at any time, which is useful if a guest doesn't pay the bill on time.

Biometric systems control access by comparing some physical aspect of a person to information on file. Hand- and finger-scanning technology are popular among biometric systems. Figure 14.1 shows a hand reader.

A popular type of magnetic key is the *Corby data chip*, which is a dime-shaped, sealed steel canister that contains sophisticated electronics to store a personal identification number (PIN). This lets it be easily attached to any smooth surface, including photo ID cards, badges, and key chains. The design also protects the electronic circuits inside the canister from dirt, moisture, corrosion, and static discharge.

Touching a data chip to the reader instantly transfers a 46-bit stream of digital data that gives the user access to a secured area. Unlike keys or security cards, the data chip is

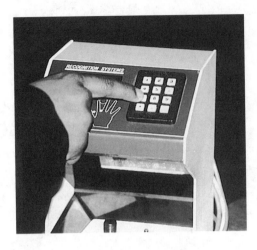

Figure 14.1   A hand reader is a popular biometric device.

user-forgiving. It doesn't need to be precisely aligned to transfer its digital data.

Along with the chip reader, you need a lock of some type. Usually, this is an electric strike or magnetic lock. (For more information on electric strikes and electromagnetic locks, see Chapters 10 and 12.) Often an intercom or telephone system is also part of the system.

An Access Control Configuration for an Apartment

Access control systems can be designed in many different ways, depending on your creativity and the needs of your customer.

Here's one configuration used for an apartment building. When someone comes to one of the main doors during the daytime, the person can enter the building by pushing a push pad (or simply by pulling the door open) because the door is unlocked. That gives the visitor access to the manager's office, but not to the apartments. The door to the apartments is kept locked all the time. At night, the main entry doors are also locked, so only people with keys can enter, such as tenants and people to whom the tenants gave a key. After entering the lobby, a guest uses the in-house phone to call a tenant and asks to be buzzed in. Because a video camera watches the lobby 24 hours a day, the tenant can turn to a certain television channel to see the person who's calling. See Figure 14.2. The tenant can then decide whether to buzz the person in.

Figure 14.2   Televisions can be used to see who's at the door.

To allow someone to "buzz open" a door, you need an electric strike. To learn more about electric strikes, read Chapter 10.

Automobile Locks

One of the most important skills a locksmith should have is the ability to open locked automobiles quickly. Although hundreds of different types of tools exist for opening vehicles, you only need a few to get into about 95 percent of them. Car-opening tools that come in a set of several dozen or more are easy to use, because they come with instructions that let you quickly match them to a specific car make, model, and year. The more you know about opening a car, the fewer tools you need. This chapter tells how to open most vehicles with a few tools.

Car opening can be a lucrative part of any locksmithing business. For some, it's their biggest source of income.

You can open most cars with five simple tools, all of which you can make yourself. Later, I tell you how to make the tools. The most important car-opening tools are a slim jim, a hooked horizontal linkage tool, an *L* tool, a *J* tool, and an across-the-car tool (or a long-reach tool). These tools have different ways of reaching and manipulating a car's lock assembly and lock buttons.

The *slim jim* is a flat piece of steel with cutouts near the bottom on both sides. The cutouts let you hook and bind a linkage rod from either side of the tool. This tool can also be used to push down on a lock pawl. Slim jims come in different widths, and it's good to have a wide one and a thin one. You can buy them at many automobile supply and hardware stores, but better models are sold through locksmith supply houses. These are often sturdier, have more notches and a handle, and they generally look more professional. To make your own, you need a 24-

inch piece of flat steel or aluminum, from 1 to 2 inches wide. You can use a ruler or other item that's the right size made from the proper material. Just draw the shape of the slim jim on to the metal, and then grind away the excess material.

Hooked horizontal linkage tools go by many names, and they come in all sizes and configurations. The small hook on the end of the tool lets you catch and bind a horizontal rod and slide it to unlock the door. Some hook down to catch the rod, while others hook from the bottom of the rod. It's good to have both of them. Two other kinds of horizontal linkage tools include the three "fingers" type that spreads to clamp onto the rod and the tooth-edged type that bites into the rod. I don't like either of those two, because when using them, you have to be especially careful to avoid bending the linkage rods.

The *J* tool is one of the easiest to use. The *J* tool goes within the door, between the window and weather stripping, and then under the window and beneath the lock button to push the button up to the unlocked position.

An *L* tool is used to push or pull on bell cranks and lock pawls. You can also use the *L* tool to access the lock rod by going under the lock handle.

For versatility, buy or make a tool that is an *L* tool on one end and a *J* tool on the other. The part of the tool that enters between the door needs to be a specific shape; the rest of the tool is the handle. Making or buying tools that have a different tool or a different-size tool on each end is useful.

The *across-the-car tool* (or *long-reach tool*) is a 6-foot (or longer) piece of 3/16-inch round stock bar with a small hook on one end. Its name comes from the fact that the tool can be used to enter a window and reach across the car to get to a lock or window button. But, sometimes, you use it on the same side of the car on which you inserted it. Most of the long-reach tools you buy come in three pieces, and you screw them together before each use. They often bend and break at the joints. If you buy one in three pieces, you should braze the pieces together, but it's best to make your own from one piece of steel.

To use car-opening tools that go into the door cavity, you also need a flexible light and one or more wedges. The wedges should be made of plastic, rubber, or wood. A newer type of wedge is called an *air wedge*. To use an air wedge, you push the wedge in place while it's deflated, and then you pump it to inflate the wedge, leaving space to insert a tool. You use the wedges to pry the door from the window or to pry one side of the door open enough to insert a car-opening tool and a flexible light. The light lets you see the linkage assembly, so you can decide what tool to use and where to place that tool.

Car-Opening Tools

With most cars, many good techniques can let you quickly open them. A locksmith who opens lots of cars tends to favor certain techniques. There's nothing wrong with that. Whatever gets you in quickly and professionally without damaging the vehicle is fine.

To open the car, parts to reach for include the lock button, the bell crank, and the horizontal and vertical linkage rods. The *bell crank* is a lever connected to the latch or other linkage rod. One popular style of bell crank is semicircular, and another style is L-shaped. A *horizontal rod*, as the name implies, runs parallel to the ground. A *vertical rod* runs vertically from the lower part of the door toward the top of the door (often to a lock button).

You don't always have to use a tool within the door. Many locks are easy to impression or pick open (for more information about lockpicking and impressioning, see Chapters 2 and 4, respectively).

When approaching an unfamiliar car model, walk around it and look though the windows. As you walk around the vehicle, consider the following:

1. Does the car have wind wings (vent windows)?
2. Does the car have a lock button at the top of the door?
3. Does the car have a lock button on the inside of the door?
4. Does the lock button move horizontal or vertical?
5. Are there any gaps around the doors where you may be able to insert an opening tool?
6. What type of linkage is used—horizontally or vertically?
7. Can you gain access to the vehicle by removing the rear view mirror?
8. Can you manipulate the lock assembly through a hole under the outside door handle?
9. What type of pawl is used? (As a rule, pre-1980 locks have free-floating pawls. Later models have rigid pawls.)

Using a J Tool

If the vehicle has a lock button on the top of the door, you may be able to open it with a J-shaped tool. First, insert a wedge between the door's weather stripping and window to give you space for the tool. Insert the *J* tool into the door until it passes below the window. Then, turn the tool so its tip is under the lock button. Lift the lock button to the open position. Carefully twist the tool back into the position in which you had inserted it and remove the tool, without jerking on it, before removing the wedge.

The Long-Reach Tool

If you learn to use it, the long-reach tool can be one of the most useful car-opening tools you have. You can quickly unlock most vehicles with it—including many of the latest models. When you use the tool, it's as if you have a very long and very skinny arm. The tool lets you reach inside a crack of a car door to push, pull, press, and rotate knobs and buttons. You can even use it to pick up a set of keys!

To use the long-reach tool, first you place a wedge near the top of the door to pry the door open enough to insert the tool (sometimes you may need to use an extra wedge). Use a protective sleeve at the opening, and then slide the tool into the sleeve. The protective sleeve is to prevent the tool from scratching the car. (You could also use cardboard or the plastic label from a bottle of soda pop.)

Most of the long-reach tools you can buy are about 56 inches long, which isn't always long enough. If you purchase a long-reach tool, get the longest one you can find. You can make your own with a 6-foot-long, 3/16-inch-diameter stainless steel rod. On one side of the rod make a 1-inch bend at a 90-degree angle. Dip that 1-inch bend into plati-dip or some other rubber-like coating (to give it a nonscratch coating).

Making Other Tools

You can find supplies at many hardware and home improvement stores to make your own car-opening tools. You need flexible flat stock and bar stock of different sizes. See Figures 15.1, 15.2, and 15.3 for patterns for making some useful tools.

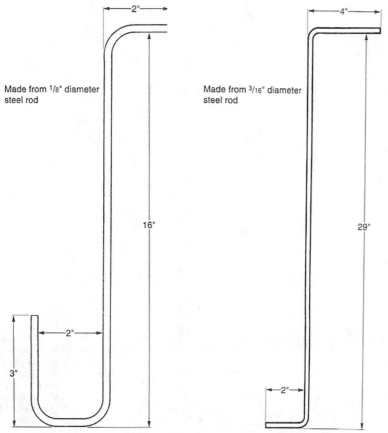

Made from ¹/₈" diameter
steel rod

Made from ³/₁₆" diameter
steel rod

Figure 15.1 An under-the-window or "J" tool is useful for raising the door lock button up to open a vehicle.

Figure 15.2 An "L" tool can be useful for pushing or pulling a lock pawl to open a vehicle.

Another tool I like can be made from plastic strapping tape that's used for shipping large boxes. It's hard to find this tape for sale in consumer outlets. I get mine free from department stores before they throw it away. Take about 2 feet of strapping tape, fold it in half, and glue a 3-inch-long piece of fine sandpaper to the center. When it dries, you have a nice stiff tool that can easily slide between car doors to loop around a lock button to pull up the button. It works like the *J* tool, but from the top of the button, instead of the bottom. The sandpaper isn't critical, but it helps the tool grab the button more easily.

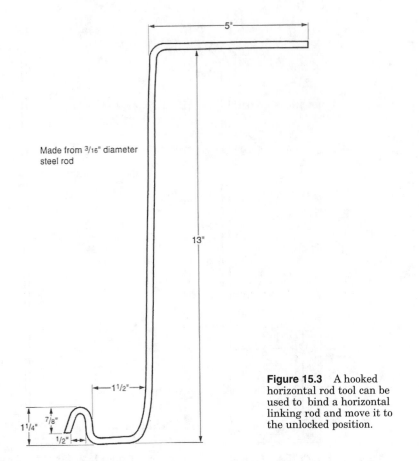

Made from ³/₁₆" diameter steel rod

Figure 15.3 A hooked horizontal rod tool can be used to bind a horizontal linking rod and move it to the unlocked position.

Business Considerations

Often a locked-out person will call several locksmiths and give the job to the one who gets there first. Or, the person may get their car door open before the locksmith gets there. Either way, you may not be able to collect a fee, unless you made it clear when you received the call that you have a minimum service charge for going on all car-opening service calls. If possible, try to get the person's debit or credit card number before you leave. Again, emphasize that you have a minimum service call that will be charged to their card.

Before you begin to work on a door, ask the customer what attempts were already made to open the door. To improve your chance of getting an honest answer, ask in a manner that sounds like you're just gathering technical information to help you work. If you learn someone has been fooling around with a

Car-Opening Dispatch Procedure

Having a good dispatch protocol can help you stay out of legal trouble, get the information you need to unlock the vehicle quickly, and make sure you get paid. Modify this protocol to fit your needs:

1. Speak directly to the owner or driver of the vehicle and not to a middle-person. If the owner or driver can't come to the telephone, don't go to the job.

2. Have the person verbally confirm they want you to do the job and that they are authorized to hire you.

3. Always quote an estimated price (or the complete price) and a minimum service-call fee. Explain that the service-call fee is for the trip and will be charged even if no other services are performed.

4. Ask how the charges will be paid (credit card, cash, or check). Explain that all charges must be paid in full and are due on your arrival.

5. Get the make, model, year, and color of the vehicle, as well as its license plate number.

6. Get the exact location of the vehicle. If the customer isn't sure, ask to speak to someone who is.

7. Get a phone number to call back on, even if it's a pay phone. Tell the person someone will call back in a moment to confirm the order.

8. Call the phone number to confirm that someone is really there. If no one answers, don't go to the job.

9. When you get to the job, ask to see identification, and make sure keys are in the car. Also, have the person sign an authorization form.

10. When you open the car, grab the keys and keep them until you've been paid. If you're given any hassles about payment, toss the keys back in the car and close the locked door.

door, don't work on that one. You don't want to be held responsible for any damage someone else may have caused.

To open a lock with vertical linkage rods, you can often use a slim jim to pull up the rod to the unlocked position or use an under-the-window tool to lift up the lock button. Before using an under-the-window tool on a tinted window, lubricate the tool with dishwashing liquid. That helps to reduce the risk of scratching off the tint. You may also be able to use an *L* tool to pull up the bell crank, which is attached to the vertical linking rod.

To use a hooked horizontal rod tool, first insert a wedge between the door frame and the weather stripping, and then lower an auto light, so you can see the linkage rods. Lower the hooked end on to the rod you need, twist the tool (thus, binding the rod), and then push or pull the rod to open the lock.

You may also want to buy a set of vent window tools for special occasions. Opening a vent (or wing) window is easy, but old weather stripping tears easily. In most cases, if a car has a vent window, it can be opened using basic vertical linkage techniques. If you decide to use the vent window, lubricate the weather stripping with soap and water at the area where you plan to insert the tool. Then, take your time and be gentle.

Special Considerations

Some models, such as the AMC Concord and Spirit, have obstructed linkage rods. Picking those locks may be best. Late-model cars can be tricky to work on. Many have lipped doors that make it hard to get a tool down into the door or they have little tolerance at the gaps where you insert wedges and tools. The tight fit makes damaging the car easy. Also, the owners may be especially watchful, concerned about any scratches you might make. To reduce the risk of scratching the car, use a tool guard to cover the tool at the point where it contacts the car.

Be careful when opening cars that have airbags. They have wires and sensors in the door. If you haphazardly jab a tool around in the door, you may damage the system. Use a wedge and a flexible light, and make sure you can see what you're doing.

Why People Call You to Open Their Cars

A lot of people know about using a slim jim. They're sold in many hardware and auto supply stores. And, many people know about pushing a wire hanger between the door and window to catch the lock button. Typically, people try those and other things before calling a locksmith. They call a locksmith because it's cold outside, late at night, raining (or all three), and they grow tired of trying to unlock their cars themselves.

Newer model cars are harder to get into using old slim-jim and wire-hanger techniques.

People seldom break their windows on purpose, even in emergencies. Replacing a car window is expensive and inconvenient, and there's a psychological barrier to smashing your own car window.

16

Masterkeying

Masterkeying can provide immediate and long-range benefits that a beginning locksmith can find desirable. This chapter covers the principles of masterkeying and techniques for developing master key systems.

Coding Systems

Coding systems help the locksmith distinguish various key cuts and tumbler arrangements. Without coding systems, masterkeying would be nearly impossible.

Most coding systems (those for disc, pin, and lever tumbler locks) are based on depth differentiation. Each key cut is coded according to its depth; likewise, each matching tumbler receives the same code. Depths for key cuts and tumblers are standardized for two reasons: standardizing these depths is more economical (mass production would be impossible without some kind of standardization) and depths, to some extent, are determined by production.

Master Key Systems

In most key coding systems, tumblers can be set to any of five possible depths. These depths are usually numbered consecutively 1 through 5. Because most locks have five tumblers— each one with five possible settings—there can be thousands of combinations. Master keys are possible because a single key can be cut to match several lock combinations.

In developing codes, certain undesirable combinations cannot be used. The variation in depths between adjoining tumblers cannot be too great. For example, a pin tumbler key cannot be cut to the combination 21919 because the cuts for the 9s rule out the cuts for the 1's. Likewise, a pin tumbler lock with the combination 99999 would be too easy to pick. The undesirable code combinations vary depending on the type of tumblers, the coding system, and the number of possible key variations. The more complex the system, the greater the possibility of undesirable combinations.

Masterkeying Warded Locks

Because a *ward* is an obstruction within a lock that keeps out certain keys not designed for the lock, a master key for warded locks must be capable of bypassing the wards. Figure 16.1 shows a variety of side ward cuts that are possible on warded keys. The *master key* (marked *M*) is cut to bypass all the wards in a lock admitting the other six keys.

As explained earlier, cuts are also made along the length of the bit of a warded key. These cuts correspond to wards in the lock. To bypass such wards, a master key must be narrowed.

Because of the limited spaces on a warded key, masterkeying is limited in the warded lock. The warded lock, because it offers only a limited degree of security, uses only the simplest of master key systems. Figure 16.2 shows some of the standard master keys available from factories.

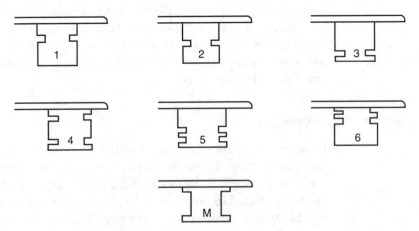

Figure 16.1   Side ward cuts.

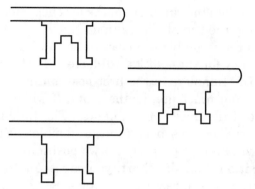

Figure 16.2 Standard factory supplied master keys.

Masterkeying Lever Tumbler Locks

Individual lever locks may be masterkeyed locally, but any system that requires a wide division of keys must be set up at the factory. A large selection of tumblers is required. The time involved in assembling a large number would make the job prohibitive for the average locksmith.

There will be occasions when you are asked to masterkey small lever locks. There are two systems. The first is the double-gate system (see Figure 16.3) and the second is the wide-gate system (see Figure 16.4). The *double-gating system* is insecure. As the number of gates in the system increases, care must be taken to prevent cross-operation between the change keys. For example, you may find a change key for one lock acting as the master key.

With either system, begin by determining the tumbler variations for the lock series in question. Otherwise, disassemble

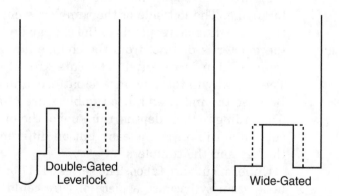

Figure 16.3 The double gate system.   **Figure 16.4** The wide gate system.

the locks and note the tumbler depths for each one. Next, make a chart listing the tumbler variations.

The master key combination can be set up fairly easily now. The master key for these locks will have a 21244 cutting code.

Suppose the tumblers in the first position have depths of 1, 2, 3, and 5. You must file depths 1 and 3 wider to allow the depth 2 cut of the master key to enter. The tumbler with the depth 5 cut requires a separate cut, or double gating. It needs a cut that will align it properly at two positions.

From position 2 on the chart, you can see that four levers must be cut; all these require a double-gating cut. In position 3, four require a double-gating cut and one needs widening. In position 5, only one requires another gating and four require widening at the current gate.

Another masterkeying method is to have what is known as a master-tumbler lever in each lever lock. The *master tumbler* has a small peg fixed to it that passes through a slot in the series tumblers. The master key raises the master tumbler. The peg, in turn, raises the individual change tumblers to the proper height, so the bolt post passes through the gate in each lever. The lock is open.

This system should be ordered from the manufacturer. The complexities of building one yourself require superhuman skill and patience.

Masterkeying Disc Tumbler Locks

Disc tumbler locks have as few as three discs and as many as twelve or more. The most popular models have five discs.

Figure 16.5 shows how the tumbler is modified for masterkeying. The left side of the tumbler is cut out for the master; the right side responds to the change key. The key used for the master is distinct from the change key in that its design configuration is reversed. The cuts are, of course, different. The key way in the plug face is patterned to accept both keys. Because the individual disc tumblers are numbered from 1 to 5 according to their depths, thinking of the master key and disc cuts on a 1 to 5 scale is easy, but on different planes for both the key and the tumblers.

Uniform cuts are taboo. The series 11111 or 22222 would give little security because a piece of wire could serve as the key. Other uniform cuts are out because they are susceptible to

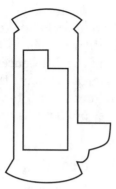

Figure 16.5   The tumbler is modified for masterkeying.

shimming. To keep the system secure, it is best to keep a two-depth interval between any two change keys. For example, 11134 is only one depth away from 11133, so 11134 should be used and 11133 omitted. The rationale is that 11134 is the more complex of the two.

The next step is to select a master key combination composed of odd numbers. At the top of your worksheet, mark the combination you selected for the master key. Below it, add a random list of possible change-key numbers. If you choose a systematic approach in developing change-key numbers, you compromise security. On the other hand, the systematic approach ensures a complete list of possible combinations. Begin systematically, and then randomly select the change-keys combinations.

A single code could be used for all customers. The main point to remember is to use different keyways.

Masterkeying Pin Tumbler Locks

Masterkeying is more involved than modifying the cylinder. It requires the addition of another pin sandwiched between a top and bottom pin. This pin, logically enough, is called the *master pin*.

The master system is limited only by the cuts allowed on a key, the number of pins, and the number of pin depths available. Because this book is for beginning and advanced students, the subject is covered on two levels: the simple master key system for no more than 40 locks in a series and the more complex system involving more than 200 individual locks.

Important to remember is that a master key system should be designed in such a way as to prevent accidental cross keying.

Pins are selected on the basis of their diameters and lengths. Master-pin lengths are built around the differences among the individual pin lengths. Consider, for a moment, the Yale five-pin cylinder with pin lengths ranging from 0 to 9 cuts. Each pin is 0.115 inch in diameter. Lower pin lengths are as follows:

0 = 0.184 inch	5 = 0.276 inch
1 = 0.203 inch	6 = 0.296 inch
2 = 0.221 inch	7 = 0.315 inch
3 = 0.240 inch	8 = 0.334 inch
4 = 0.258 inch	9 = 0.393 inch

As an illustration, let's masterkey ten locks with five tumblers each. Each tumbler can have any of ten different individual depths in the chamber.

Determine the lengths of each pin in each cylinder. Mark these down on your worksheet. The master key selected for this system may have one or more cuts identical to the change keys.

Cut a master key to the required depths. In this instance, load each cylinder plug by hand.

Using the known master key depth, subtract the depth of the change key from it. The difference is the length of the master key pin. If the change key is 46794 and the master key is 68495, the master pin combination is as follows:

Chamber Position	Bottom	Master
1	42	
2	62	
3	73	
4	90	
5	41	

Follow this procedure for each plug. Notice that not all the chambers have a master pin. Such complexity is unnecessary and makes the lock more vulnerable to picking. Each master pin represents another opportunity to align the pins with the shear line.

In practice, locksmiths avoid most of this arithmetic by compensating as they go along. For example, an unmasterkeyed cylinder has double pin sets. Masterkeying means an additional pin is added in (usually) position 1. This drives the bottom pin lower into the keyway. The bottom pin must be shortened to compensate.

A grand master key or a great-grand master key adds complexity to the system. You have two choices: add master pins in adjacent chambers or stack pins in the first chamber. Suppose you have a No. 4 bottom pin and the appropriate master pin. To use grand and great-grand master keys, you must add two No. 2 pins, so all four pins will operate the lock.

Developing the Master Key System

A master key system should be planned to give the customer the best security the hardware is capable of. Begin by asking the customer these questions:

- Do you want a straight master key system with one master to open all locks? Or, do you want a system with submasters? That is, do you want a system with several submasters of limited utility and a grand or great-grand master?
- What type of organizational structure is within the business? Who should have access to the various levels?
- How many locks will be in each submaster grouping?
- Is the system to be integrated into an existing system, or will the system be developed from scratch?
- What type of locks do you have?

Once you have considered these answers (with the help of your catalogs, specification sheets, technical bulletins, and experience), you are ready to begin developing a master key system.

Standard Key Symbol Code

Great-grand master keys are identified by the letters GGM. Grand masters carry a single letter, beginning with the first in the alphabet and identifying the hierarchy of locks the individual grand master keys open. Master keys carry two letters:

the first identifies its grand master, the second identifies the series of locks under it. Thus, a master key labeled AA is in grand master series A and opens locks in master key series *A*. Master key AB is under the same grand master, but opens locks in series *B*. Master BA is under grand master *B* and opens locks in series *A*. Change keys are identified by their master key and carry numerical suffixes to show the particular lock they open. Any key in the series can be traced up and down in the hierarchy. Thus, if you misplace change key AB4, you know that master key AB, grand master A, or the great-grand master will open the lock.

There are special keys that are out of series. Some of these keys are mentioned in the glossary, together with the appropriate key symbols for them.

If cross keying is introduced into the system—that is, if a key can open other locks on its level—the cylinder symbol should be prefixed with an *X*. If the cylinder has its own key, it is identified with the standard suffix. For example, XAA4 is a change-key cylinder that is fourth on this level. It may be cross keyed with AA3 or any other cylinder or cylinders on this level. By the same token, master key AA, grand master *A*, and the great-grand master will open it. Elevator cylinders are often cross keyed without having an individual change key. It is no advantage to have a key that will operate the elevator cylinder and no other in the system. These cylinders are identified as XIX, X2X, and so on.

The symbols that involve cross keying apply to the cylinder only; all other symbols apply to the cylinder and the key. This point may seem esoteric, but ignoring it causes the factory and everybody else grief. For example, there is neither an XIX key nor an XAA4 key. Change key AA4 fits cylinder XAA4 that happens to be cross keyed with another cylinder. Key AA4 does not fit any cylinder except XAA4. Certain advantages exist to using the standard key symbol code:

- It is a standardized method for setting up the keying systems.

- It maintains continuity from one order to the next.

- It indicates the position of each key and each cylinder in the hierarchy.

- It helps to control cross keying because each cross-keyed cylinder is clearly marked.

- It offers a method of projecting future keying requirements.
- It can be easily rendered on a chart.
- It allows better control of the individual keys within the system.
- It is a simple method of selling and explaining the keying system to an architect or building owner.

More About Masterkeying Pin Tumbler Locks

Almost any pin tumbler lock can be masterkeyed, by inserting additional tumblers called "master pins" between the drivers and key pins. The *master pins* create additional shear lines, which allow one or more additional keys to operate the lock.

The smallest master key system would include two locks and two keys. One key might open the main entry door, and the other would open the main entry door and the door to the boss's office. Using the standard key coding system, such a system would read as follows:

Door 1   (Hardware Description)   Keyed AA

Door 2   (Hardware Description)   Keyed 1AA

A larger master key system might include an office supply room. The boss wants their key to operate all three locks, but wants to prevent other employees from getting into the supply office. Your key System Schematic would look as follows:

Door 1   (Hardware Description)   Keyed AA

Door 2   (Hardware Description)   Keyed 1AA

Door 3   (Hardware Description)   Keyed 2AA

In a larger system, the boss may want a key that opens every lock, two employees who can open the front and back entry door, and a secretary who can open the front entry door and the office supply room. The schematic for such a system would be as follows:

Door 1 (Boss's Office)   (Hardware Description)   1AA

Door 2 (Employee 1 Office) (Hardware Description)   2AA

Door 3 (Employee 2 Office) (Hardware Description)   3AA

Door 4 (Office Supplies)	(Hardware Description)	4AA
Door 5 (Front Door)	(Hardware Description)	5AA
Door 6 (Back Door)	(Hardware Description)	5AA

Usually a series of locks are designed to operate by the same master key. In a large master key system, there will be submaster keys that open a series of locks within a master system, but not all of the locks. A larger system may include grand master and great-grand master keys.

A *change key* is a key that operates a single lock within the system. For instance, in a hotel, a guest may have a change key that only operates the door to their room, but that doesn't open any other guest doors.

The *submaster* key operates locks within a single area or group. For instance, a cleaning person may have a submaster key that opens all the doors on a single floor.

The *floor master key* operates one or more submaster locks. For instance, a security person may have a key that opens all the guests' doors, and perimeter doors that lead to tenant spaces.

The *grand master key* operates one or more master systems. The *great-grand master key* operates one or more grand master systems.

17

Safe Basics:
Buying and Selling

Virtually everyone has documents, keepsakes, collections, or other valuables that need protection from fire and theft. But, most people don't know how to choose a container that meets their protection needs, and they won't get much help from salespeople at department stores or home improvement centers. By knowing the strengths and weaknesses of various types of safes, you can have a competitive edge over such stores.

No one is in a better position than a knowledgeable locksmith to make money selling safes. Little initial stock is needed. They require little floor space. And safes allow for healthy price markups. This chapter provides the information you need to begin selling safes to businesses and homeowners.

Types of Safes

There are two basic types of safes: fire (or "record") and burglary (or "money"). *Fire safes* are designed primarily to safeguard their contents from fire, and *burglary safes* are designed primarily to safeguard their contents from burglary. Few low-cost models offer strong protection against both hazards. This is because the type of construction that makes a safe fire-resistant—thin metal walls with insulating material sandwiched in between—makes a safe vulnerable to forcible attacks. And, the construction that offers strong resistance to attacks—thick metal walls—causes the safe's interior to heat up quickly during a fire.

Most fire/burglary safes are basically two safes combined—usually a burglary safe installed in a fire safe. Such safes can be expensive. If a customer needs a lot of burglary and fire protection, you might suggest they buy two safes. To decide which type of safe to recommend, you need to know what your customer plans to store in it.

Safe Styles

Fire and burglary safes come in three basic types, based on where the safe is designed to be installed. The styles are wall, floor, and in-floor. *Wall safes* are easy to install and provide convenient storage space. Unless installed in a brick or concrete block wall, such safes generally provide little burglary protection. When installed in a drywall cutout in a home, regardless of how strong the safe is, a burglar can just yank the safe out of the wall and take it with him.

Floor safes (Figure 17.1) are designed to sit on top of a floor. (Some locksmiths refer to an in-floor safe as a floor safe.) Burglary models should either be over 750 pounds or bolted into place. One way to secure a floor safe is to place it in a corner, and then bolt it to two walls and to the floor. (If you sell a large safe, make sure your customer knows the wheels should be removed.)

In-floor safes (Figure 17.2) are designed to be installed below the surface of a floor. Although they don't meet construction guidelines to earn a UL fire rating, properly installed in-floor safes offer a lot of protection against fire and burglary. Because fire rises, a

Figure 17.1   Floor safes (courtesy of Gardall Safe Corporation).

Figure 17.2   In-floor safes come in many sizes (courtesy of Gardall Safe Corporation).

safe below a basement floor won't quickly get hot inside. For maximum burglary protection, the safe should be installed in a concrete basement floor, preferable near a corner. That placement makes it uncomfortable for a burglar to attack the safe.

Make sure your customer knows they should tell as few people as possible about their safe. The fewer people who know about a safe, the more security the safe provides.

Depository safes are used by businesses. See Figure 17.3. Such safes have a slot in them, so cashiers can insert money

Figure 17.3   Depository safes are used by businesses (courtesy of Gardall Safe Corporation).

into them. The slot prevents people from taking the money out again without using a key or combination.

Installing an In-Floor Safe

Although procedures differ among manufacturers, most in-floor safes can be installed in an existing concrete floor in the following way:

1. Remove the door from the safe, and tape the dust cover over the safe opening.

2. At the location where you plan to install the safe, draw the shape of the body of the safe, allowing 4 inches of extra width on each side. For a square body safe, for example, the drawing should be square, regardless of the shape of the safe's door.

3. Use a jack hammer or a hammer drill to cut along your marking.

4. Remove the broken concrete and use a shovel to make the hole about 4 inches deeper than the height of the safe.

5. Line up the hole with plastic sheeting or a weatherproof sealant to resist moisture buildup in the safe.

6. Pour a 2-inch layer of concrete in the hole, and then level the concrete to give the safe a stable base to sit on.

7. Place the safe in the center of the hole and shim it to the desired height.

8. Fill the hole with concrete all around the safe and use a trowel to level the concrete with the floor. Allow 48 hours for the concrete to dry.

9. After the concrete dries, trim away the plastic and remove any excess concrete.

See Figure 17.4. In-floor safes are usually installed near walls to make it hard for burglars to use tools on them.

Moving Safes

Getting a heavy safe to your customer can be backbreaking unless you plan ahead. Consider having the safe drop-shipped, if that's an option. Most suppliers will do that for you.

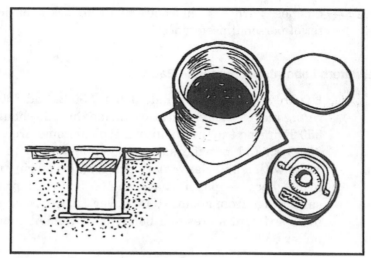

Figure 17.4   Installing an in-floor safe in concrete.

As a rule of thumb, have one person help for each 500 pounds being moved. If a safe weighs more than a ton, however, use a pallet jack or a machinery mover. When moving a safe, never put your fingers under it. If the safe has a flat bottom, put three or more 3-foot lengths of solid steel rods under it to help slide the safe around, and use a pry bar for leverage.

Special Safe Features

Important features of some fire and burglary safes include relocking devices, hardplate, and locks. Relocking devices and hardplate are useful for a fire safe, but they are critical for a burglary safe. If a burglar attacks the safe and breaks one lock, the *relocking devices* automatically move into place to hold the safe door closed. *Hardplate* is a reinforcing material strategically located to hinder attempts to drill the safe open. Never recommend a burglary safe that doesn't have relockers and hardplate.

Safe locks come in three styles: key-operated, combination dial, and electronic. The most common, *combination dial* models, are rotated clockwise and counterclockwise to specific positions. *Electronic* locks are easy to operate and provide quick access to the safe's contents. Such locks run on batteries that must be recharged occasionally. For most residential and small business purposes, the choice of a safe lock is basically a matter of personal preference.

Underwriters Laboratories Fire Safe Ratings

UL fire safe ratings include 350-1, 350-2, and 350-4. A *350-1 rating* means the temperature inside the safe shouldn't exceed 350°F during the first hour of a typical home fire. A safe rated 350-2 should provide such protection for up to two hours. Safes with a 350-class rating are good for storing paper documents because paper chars at 405°F. Retail prices for fire safes range from about $100 to over $4,000. Most models sold in department stores and home improvement centers sell for under $300.

Underwriters Laboratories Burglary Safe Standard

The UL 689 standard is for burglary-resistant safes. Classifications under the standard, from lowest to highest, include Deposit Safe, TL-15, TRTL-15x6, TL-30, TRTL-30, TRTL-30x6, TRTL-60, and TXTL-60. The classifications are easy to remember when you understand what the sets of letters and numbers mean. The two-set letters in a classification (TL, TR, and TX) signify the type of attack tests a safe model must pass. The first two numbers after a hyphen represent the minimum amount of time the model must be able to withstand the attack. An additional letter and number (for example, x6) tells how many sides of the safe have to be tested.

The *TL* in a classification means a safe must offer protection against entry by common mechanical and electrical tools, such as chisels, punches, wrenches, screwdrivers, pliers, hammers and sledges (up to 8-pound size), and pry bars and ripping tools (not to exceed 5 feet in length). *TR* means the safe must also protect against cutting torches.

TX means the safe is designed to protect against cutting torches and explosives.

For a safe model to earn a *TL-15* classification, for example, a sample safe must withstand an attack by a safe expert using common mechanical and electrical tools for at least 15 minutes. A *TRTL-60 safe* must stand up to an attack by an expert using common mechanical and electrical tools, as well as cutting torches, for at least 60 minutes. A *TXTL-60 safe* must stand up to an attack with common mechanical and electrical tools, cutting torches, and high explosives for at least 60 minutes.

In addition to passing an attack test, a safe must meet specific construction criteria before earning an Underwriters Laboratories (UL) burglary safe classification. To classify as a deposit safe, for example, the safe must have a slot or otherwise provide a means for depositing envelopes and bags containing currency, checks, coins, and the like into the body of the safe. And, it must provide protection against common mechanical and electrical tools.

The TL-15, TRTL-15, and TRTL-30 safe must either weigh at least 750 pounds or be equipped with anchors and instructions for anchoring the safe in a larger safe, in concrete blocks, or to the premises in which the safe is located. The metal in the body must be the equivalent to solid open-hearth steel at least 1-inch thick, having an ultimate tensile strength of 50,000 pounds per square inch (psi). The TRTL-15x6, TRTL-30x6, and TRTL-60 must weigh at least 750 pounds, and the clearance between the door and jamb must not exceed 0.006 inch. A TXTL-60 safe must weigh at least 1,000 pounds.

TL-15 and TL-30 ratings are the most popular for business uses. Depending on the value of the contents, however, a higher rating may be more appropriate. Price is the reason few companies buy higher-rated safes. The retail price of a TL-30 can exceed $3,500. A TXTL-60 can retail for over $18,000.

Such prices cause most homeowners and many small businesses to choose safes that don't have a UL burglary rating. When recommending a nonrated safe, consider the safe's construction, materials, and thickness of door and walls. Better safes are made of steel and composite structures (such as concrete mixed with stones and steel). Safe walls should be at least 1/2-inch thick and the door at least 1-inch thick. Make sure the safe's bolt work and locking mechanisms provide strong resistance to drills.

Selling More Safes

You'd have to cut a lot of keys to make the money you can make from selling and installing a safe. No one is in a better position than locksmiths to sell high-quality safes. Whether you're just starting to sell them or you've been selling them for years, you can boost your sales.

The key to selling more safes is for you and all your sales-persons to focus on the four Ps of marketing: products, pricing, promotion, and physical distribution.

Selling More Products

One of the most important marketing decisions you can make is which safes to stock and recommend. You need to consider quality, appearance, cost, warranty, and delivery time. Only sell good safes you believe in. Your enthusiasm for the safes you sell can make it easier for you to talk about them.

Little initial stock is needed to start selling safes. If you want to be taken seriously, however, you need to have a few on display. Most people want to see and touch a safe before buy-ing it—much like when buying a car. Stock several sizes of each type of safe. This can make it easier for you to sell the cus-tomer up to a more expensive model.

If you're just starting to sell safes, don't stock large, heavy models because they're expensive, hard to transport, and usu-ally don't sell quickly. Consider stocking floor fire safes and in-floor safes. In some locales, you also may want to stock gun safes. If you're planning to sell to businesses, stock TL-rated floor safes and depository safes. Square-door safes usually sell faster than round-door models.

In addition to choosing which products to sell, you need to choose a distributor. Some distributors have a "safe dating pro-gram," in which they let you stock safes without having to pay for them until you sell them. If your distributor doesn't offer such a program, ask about stocking the safes for 90 days before paying. And see if the distributor offers training seminars.

You and your salespeople must become familiar with what you're selling. Study literature about the safes, and attend dis-tributor and manufacturer seminars. If you don't know much about your products, potential customers will notice. As a lock-smith, you're selling your expertise as well as safes. If price were the only factor, people would buy low-end safes from department stores and home-improvement centers instead of high-quality safes from you. Major manufacturers regularly offer seminars on installation and sales.

Pricing

Buy safes at good prices and sell them at a reasonable markup. Don't worry about not having the lowest prices in town. Marketing for peak profits involves adjusting the prices of products to meet the needs of customers and the needs of your company. Adjustments in prices mean adjustments in the customer's perception of prices. Prices should be based on perceived value. Many customers choose a more expensive product because they believe in the adage, "You get what you pay for."

Some customers always refuse to pay the sticker price; they feel better if they dicker the price down. You need to price your safes so all types of customers feel they're getting a good deal. One way to do this is to price most items with a little room for dickering. A good idea is to price so you can negotiate slightly— such as by giving a 2 percent discount for cash payment.

Be careful about lowering your prices, however. Every attempt should be made to sell the safe at sticker price. If the customer objects, point out the safe's benefits and features. Remember, you're selling a specialty product that protects the valuables and keepsakes of a family or business. Make it clear that you're selling a high-quality product. One major safe dealer emphasizes the importance of quality by displaying a cheap fire safe that had been broken into.

Promotion

The quality of your promotional efforts has a lot to do with how much money a customer is willing to pay for your safes. Promotion is mainly in the form of imaging and advertising. To design an effective imaging plan, you need to consider everything your customers see, hear, and smell during and after the selling process. Pay close attention to detail.

Your showroom needs to be pleasant for customers. The safes should be displayed where they can be readily seen and touched, but not where customers could trip over them. Your customers should have to walk by safes whenever they come into your shop. The display area should have good lighting, be clean (don't let dust build on the safes), and be at a comfortable room temperature. Use racks and elevated platforms, so customers don't have to bend down to touch your smaller safes.

Use plenty of manufacturer posters, window decals, and brochures in the display area. Such materials help to educate customers about your safes. Some safe manufacturers offer display materials.

In addition to placing promotional literature near the safes, include a product label on each safe. The label should include the following information: safe brand, rating, special features, warranty, regular price, sale price (if any), and delivery and installation cost. This information can help you better describe the product to customers.

A lot of locksmith shops have a web site. This can help you make direct sales, as well as promote all your products and services. You can even include a map to make it easier for customers to find you. The key to having a successful web site is to get an easily remembered domain name. If it isn't already being used on the Internet, you can use your shop name. To find out if a domain name is in use, go to www.networksolutions.com.

After getting a domain name, you can use one of the many web-site creation programs to make your web site or you can hire someone to do it. Expect to pay at least a few hundred dollars for someone to make a basic web site for you. To get ideas for creating a web site, go to an Internet search engine, such as www.hot-bot.com, and enter *lock and safe* or *safe and lock*. You'll find many locksmiths' web sites. Once you have a web site, you need to promote it by including your web address on your letterhead, business cards, service vans, and Yellow Pages ads.

The most important advertisement you can have is a listing in your local Yellow Pages. When people are looking for a safe, they don't read the newspaper, they reach for the phone book. Consider a listing under "Locks and Locksmithing" and "Safes and Vaults."

The larger your ad, the more prominent your company seems (and the more costly the ad is going to be). To determine the right size ad to get, look at those of your competitors. If no one else has a display ad, for instance, then don't get a display ad for that Yellow Pages heading. Instead, consider getting a bold-type listing. If many of your competitors have full- and half-page ads, however, you should have one, too—if you can afford it. If not, get the largest ad you can afford.

Some large safe dealers find television ads useful. Although advertising on national television can be expensive, advertising on local and cable television can be cost-effective. To do successful television advertising, you need to create professional-quality commercials and run them regularly. You can't simply run commercials for a month or two and expect long-term results.

The most successful safe dealers take every opportunity to talk up their safes. Whenever someone buys something at your store, ask if they need a safe. And, ask during every service call. Be prepared to talk about the benefits of buying one of your safes—convenience, protection, and peace of mind. Explain that you install and service your safes. Even if the person isn't ready to buy one now, they may remember you when they are ready to buy.

Physical Distribution

The sale of a safe isn't the end of the transaction. Delivery of a safe should be done as soon as possible after the sale. Slight paranoia is a natural symptom in a customer who has just purchased a safe. If the safe takes too long to be delivered, the customer may want to cancel the order. Only work with distributors who stock a lot of safes and who can get them to you quickly.

Delivery should be done professionally and discreetly. Some safe retailers use unmarked vehicles to deliver safes. If you use an unmarked vehicle, be sure to point that out to the potential customer when you're trying to sell the safe.

By taking a little time to evaluate your current marketing strategy, you can find ways to make it better. Just remember to carefully coordinate your decisions about the four Ps of marketing and you'll improve the fifth P—Profits.

18

Drilling and Manipulating Safes

Opening a safe by drilling should be your last resort. When you do need to drill, strive to do so with only one small hole. Damaging a safe is easy, as is turning a small problem into a big one. If you've never drilled a safe open, work with a professional locksmith or safe technician to drill open a few safes before you try to do it yourself. Or, you can buy a few used safes, and then practice drilling them open and repairing the holes.

When you're ready to drill open a safe, you need to have confidence in your ability and information about the safe. If you're not confident or you don't have good information, you're likely to drill several holes before opening the safe. You want to open safes by drilling one small hole.

To learn about safes, read all you can about them, and take classes when you can. Whenever you work on a safe, take as many photographs as you can. Develop a safe information file that includes photographs, drawings, and technical information. And buy as many safe books as you can. Then, when you need to drill a safe, you'll have information to start from.

Also, attend trade shows to meet other safe dealers and manufacturers. These people may be helpful when you have a question about safe work. Ask other safe dealers and locksmiths about tools they like to use, such as which are their favorite drill rigs, scopes, and drill templates.

Don't make the mistake of guessing where to drill. If you don't have good information about the safe, either call the manufacturer or go online to ask other locksmiths and safe technicians for help. Good web sites to go to for help include ClearStar.com, Locksmithing.com, and TheNationalLocksmith.com.

Templates are useful. You can place them on a safe door to mark the right place to drill. Drilling at the right place is 95 percent of the safe-opening job. Drilling at the wrong place is not only a waste of time, but it also causes you to drill several holes, which could damage the safe.

Safe Lock History

During the 1930s, most safe lock manufacturers standardized the size of the lock cases to the present-day dimensions. Mosler was the exception. Mosler's case size didn't become standardized until the 1970s, when they introduced the 302 series model.

Today two leading safe manufacturers—Mosler and Diebolt—produce their own locks. Whenever you see a Mosler safe, it will have a Mosler safe lock. And, whenever you see a Diebolt safe, it will have a Diebolt lock—with the exception of the G.S.A. Class 6 type government containers built by Mosler or Diebolt. Some of these safes are equipped with Sargent & Greenleaf locks because neither Mosler nor Diebolt had a safe lock that met government specifications for some time.

The Combination Lock

The combination lock isn't a complicated device. Underwriters Laboratories (UL) requires a lock to have 1 million possible combinations—which makes the lock seem complicated. If the combination has been lost, it could take a long time to guess it.

The Lock Case

UL-listed combination locks are made of a corrosive resistant material, such as a Zamac type composition—an alloy metal of aluminum and zinc. Pre–UL listed requirements lock cases were made of brass, bronze, steel, and cast iron. Some old safes don't have a case, but they have their wheel packs hung on the inside of the door.

The case size among safe lock manufacturers is virtually the same size. The hole pattern for the mounting screws allow retrofitting from one lock to another.

The case holds all the lock's parts, except for the dial and dial ring. The *wheel post*, in which the wheels rotate, is an integral part of the case. A retainer ring or clip secures the wheels of the lock to the wheel post.

The Lockbolt

In most cases, the *lockbolt* is made of brass. An exception is *La Gard*, which uses a bolt made of casted Zamac with Teflon.

How Big Should the Hole Be?

If you're a good safe technician, you should be able to open most safes with a single 1/4-inch hole. But, when drilling a hole over 3 inches deep into a door, you would have to drill precisely to hit your target. Also, a thin drill bit is likely to break when drilling into hardplate.

To drill a safe professionally, you need a wide range of drill bits of varying lengths, metal-cutting hole saws, a heavy duty drill, a drill rig, at least one thin probe, and several different size scopes. Different locksmiths use different size holes, starting at 1/4 inch. Mosler Safe Co. offers a drill rig kit with four drill-bit sizes—from 3/8 inch to 3/4 inch. If you're using a scope, use the smallest hole that allows the barrel of your scope to fit into it. If you aren't using a scope, drill the hole 3/8 inches.

To drill holes, use good metal-cutting drill bits and metal-cutting hole saws. If you face a badly burglarized safe with its hinges broken off and relocked fired, for instance, it may not be worth your time trying to drill a small hole. It may be advantageous to drill one large hole.

Where to Drill

Where you want to drill a safe depends on the type, position, and material of its locks and relockers, and whether the safe has hardplate. Although most safes are drilled from the front, it's often a good idea to drill at the top or a side of the safe. When you reach the lock case, you can use a thin probe to lift the lock bolt or manipulate the lock parts into the open position.

In addition to buying safe books, read all the articles you can on safes, and join a safe technician organization, such as the National Safeman's Organization and the Safe & Vault Technicians Association. Also, meet with other safe technicians and share information. You can meet other safe technicians online at Locksmithing.com and ClearStar.com. If you mention you're a reader of *Master Locksmithing*, by Bill Phillips, you'll get a $10 discount for your initial ClearStar.com membership.

Drill Locations

The following are the most-often-used drill locations:

- Lever Screw. To remove the lever screw from the lockbolt.
- Fence. To see and align the wheel gates under the lever fence.
- Handle Cam. To punch the cam inward to bypass the lock bolt.
- Relock Trigger. To disengage the bolt.
- Dial Ring. To view the wheels with a scope and align the gate to move to the drop area.
- Change Key Hole. To see and decipher wheels through the change key hole.
- Lock Case, Top. To see the fence and wheel gates from the top.
- Lock Case, Bottom. To see the fence and wheel gates from the bottom of the lock.
- Lock Case, End. To see the wheel gates through the end of the case.

Drilling the Lever Screw

When you aren't able to move the drive cam, such as when there is a broken or missing spline, you may need to drill out the lever screw. By drilling out the lever screw, you can then use a probe to move the bolt to the unlocked position.

Drilling the Fence

Drilling the fence is the most frequently used place to drill a safe, because it's all straight drilling. It doesn't require drilling at angles, and no special scopes are needed.

To drill for the fence, you need to remove or reduce the dial in diameter. All drilling takes place under the area usually covered by the dial. To remove the dial, push a screwdriver under the lip of the dial and gently pry off the dial. Reducing the dial diameter lets you see and align (using a thin probe) its gates under the lever fence.

Another option for removing the dial is to use an auto-body dent puller, also called a *slam hammer*. You drill a hole through the center of the dial, and then insert the hardened sheet metal screw and thread it into the drilled hole. You then slam the sliding weight on the puller shaft against the handle, pulling the dial from the spindle.

Don't pull the dial is written on a TL-rated safe, the dial of an S&G 8500 series lock, or any dial made of brass, bronze, or steel.

In some instances, it's not a good idea to pull a dial off. But, you still want to drill under the dial. In that case, you can reduce the diameter of the dial by using a 1-1/2-inch hole saw. Before cutting the dial, rotate it at least four revolutions to the right.

Drill Points

A *drill point* is a place on a safe that can be drilled to unlock the safe. Most safes have multiple drill points. I have a long list of drill points, but it's unethical to give them to people who aren't locksmiths or safe technicians. To receive a list of drill points, complete the Registered Professional Locksmith Test (see Appendix F) or the Registered Security Professional Test.

Safe Manipulation

You've probably seen movies and television shows where the thief puts his ear to the safe while turning the combination dial, and the safe opens within minutes. That's a simplified form of manipulation. In real life, however, the art of manipulation is more complicated. Manipulation involves touch, sound, and feel.

The advantage of manipulation over drilling is that you don't have to patch up holes or paint the safe.

When the right combination is dialed, the fence and lever fall into the notches, allowing the bolt to retract. If the wrong combination is dialed, the lever and fence sit on the wheels, preventing the bolt from being retracted.

The first step in manipulating a safe is to find out the drop in area. The drop in point is where the sloped notch in the drive cam contacts the lever. To find it, rotate the dial four or more times to the left to pick up all the wheels, listening to the nose of the lever. Keep turning left until the nose of the lever slightly drops into the drive cam. Keep slowly rotating the dial to the left until you get to the next indication—which will be the nose of the lever striking the right side of the drive cam. These are called *contact points*. (The contact area is the space between the contact points).

Determining the number of wheels in the lock is the next step in manipulation. Start by turning the dial four or more times to pick up all the wheels. You already have the contact area. Let's say on the dial it's between 20 and 30. Continue turning the wheel pack to the left, and park at a number far outside the contact area, such as 70.

Next dial the wheel to the right. When you pass 70 (or wherever you parked the dial), you should hear the drive pin contact the fly of the first wheel. Then you'll have the first number. Keep rotating to the right and whenever you turn to 70, you'll hear and feel another wheel being picked up. Continue the process until you don't hear anymore fly contact at 70. Then you've run out of wheels. Safes typically have three or four wheels.

The last step of manipulation is charting the data from the combination dial. Use a graph to enter left and right contact points.

To make manipulation easier, you can use an audio amplifier to help you hear what's happening in the lock. Some locksmiths use automatic combination dialers, which automatically dial all possible numbers. The problems with such dialers is that they're expensive and can take a long time to find the combination.

To become proficient in safe lock manipulation, you have to practice. You can use a cutaway safe lock.

Working as a Locksmith

In this chapter, I tell you how to find a job as a locksmith. Every month the locksmithing trade journals have dozens of help-wanted advertisements and many businesses-for-sale advertisements. The Internet also has a lot of both.

Many moneymaking opportunities are available for a person who has locksmithing skills or who wants to learn the trade. You need to decide where you want to work, seek out meetings with prospective employers, meet with them, and negotiate wages and other working terms. You also need to make decisions about joining locksmithing associations and getting bonded, certified, and licensed.

Joining a Locksmith Association

Belonging to a locksmithing association can help to improve your credibility among other locksmiths and in your community. Most of the associations offer trade journals, technical bulletins, certifications, classes, discounts on books and supplies, locksmith bonds, insurance, and so on. They provide membership certificates that you can display in your shop and logos that you can use in your advertisements. It also can be helpful to include association memberships on your résumé or in your Yellow Pages advertisement.

Many successful locksmiths, however, don't belong to any locksmithing association. Some don't join because they believe that the membership dues are too high. Others refuse to join because they disagree with the policies or legislative activities of the organizations. The criteria for becoming a member differ among associations, as do the policies, legislative activities, and membership dues. You need to decide for yourself if joining one will be beneficial for you. To join the International Association of Home Safety and Security Profesionals, complete the application in Appendix G. If you join the association, you can take the Registered Professional Locksmith test at no charge.

Bonding

Bonding is a type of insurance that pays off if you are convicted of stealing from your customers. It's a good idea to place your bonding certificate in your shop where customers can see it. The easiest way to get bonded is through one of the major locksmithing trade journals—such as the *National Locksmith*. Subscribers may opt to be bonded for a small fee. The National Automobile Association and the National Safeman's Organization provide bonding to their members as part of their membership fees.

A locksmith certification signifies that a person has demonstrated a level of knowledge or proficiency that meets a school or association's criteria for being certified. The significance of a certification depends on the integrity of the organization issuing it. Some of the best known certifications are granted by the Associated Locksmiths of America (ALOA), the American Society of Industrial Security (ASIS), the National Safeman's Organization (NSO), and the Safe and Vault Technicians Association (SAVTA). It's also helpful to be registered as a locksmith with the International Association of Home Safety and Security Professionals (IAHSSP). A registration test is provided in Appendix F. By passing that test, you can qualify for the Registered Professional Locksmith designation, and after qualifying, you may access the association's web site and include the initials RPL after your name.

Locksmiths are divided about the need for certification. Some of the most experienced and well-known locksmiths and safe technicians don't use certification initials after their names. While writing this chapter, I looked at the mastheads of the two most popular locksmithing trade journals and counted 30 regular writers between them. Of the 30, only 5 used certification initials.

Many locksmiths feel they shouldn't have to meet the approval of *a* specific school or organization. They prefer to allow the free market to determine the competency of a locksmith. An incompetent locksmith, they argue, won't be able to compete successfully against a competent one for long.

Others feel locksmith certifications can benefit the locksmith industry by raising the competency level of all locksmiths and by improving the public perception of locksmiths. In any case, certifications can be helpful in finding work, especially if you don't have a lot of experience. They show prospective employers that you're serious about wanting to learn locksmithing.

A license differs from a certification in that a *license* is issued by a municipality rather than by a school or an association. Most cities and states don't offer or require locksmith licenses. In those places, locksmiths are required only to abide by laws that apply to all businesses—zoning, taxes, building codes, and so on. (However, because the alarm industry is heavily regulated, a locksmith who wants to install alarms first will probably need to obtain an alarm systems installer's license.)

A few places, such as New York City, Texas, Tennessee, North Carolina, and California, require a person to get a locksmith license before offering locksmithing services. Contact your city, county, and state licensing bureaus to find out if you need to have a locksmith (or other) license. Some places levy stiff fines on people who practice locksmithing without a license.

The requirement for obtaining a locksmithing license varies from place-to-place. Some require an applicant to take a competency test, be fingerprinted, provide photographs, submit to a background check, and pay an annual fee of $200 or more. Other places simply require applicants to register their name and address, and pay a few dollars annually.

Locksmiths are divided on three major issues concerning licensing: whether licensing is necessary, what criteria should be required to obtain a license, and whether a national locksmith association should play an integral part in qualifying locksmiths for licensing.

Some locksmiths feel licensing is needed to protect consumers from unskilled locksmiths and to improve the image of the profession. Other locksmiths feel the free market does a good job of separating good locksmiths from the bad.

Proposed criteria for licensing—such as the fees, liability insurance, and competency testing—are also of concern to many locksmiths. The primary arguments about such matters center around whether a specific criterion is unnecessary, too costly, or too burdensome for locksmiths.

Some locksmithing trade associations are actively working to enact new locksmithing laws throughout the United States. But, only a minority of locksmiths belongs to associations. Most locksmiths are unaware of such bills until they become law. If you're a locksmith or plan to become one, you should take an interest in the current legislation related to locksmiths in your area.

Write or call your local, county, state, and federal representatives and ask to be kept informed of proposed legislation related to locksmiths. When licensing bills are proposed, ask your representative to mail a copy to you. In addition to letting your representatives know what you think of the bills, you can send letters to the editors of locksmithing trade journals.

Another great way to have your views heard by people who care is to place messages on locksmithing web sites. Some of the most popular are alt.lock-smithing (newsgroup), www.TheNationalLocksmith.com, www.Locksmithing.com, and www.ClearStar.com. By telling Jay, the ClearStar webmaster, that you learned of his web site from this book, he will take $10 off of your annual fee for the site. You can also stay current with locksmithing issues by joining the IAHSSP (see Appendix G).

Planning Your Job Search

Your job search will be most productive if you approach it in a professional manner. You need to plan every step of the way. When you're seeking a job, you're a salesperson—you're selling yourself. You need to convince a prospective employer to hire you, whether or not the employer is currently seeking a new employee.

Deciding Where You Want to Work

Before contacting prospective employers, decide what type of shop you want to work in and where you want to work. Which cities are you willing to work in? Do you prefer to work in a

large, midsize, or small shop? Do you prefer to work for a highly experienced shop owner or for an owner who knows little
about locksmithing?

You can answer these questions by considering your reasons
for seeking employment. If you're mainly looking for an immediate, short-term source of income, for example, your answers
might be different than if you were more concerned about job
security or gaining useful locksmithing experience.

In a small shop, you might be given a lot of responsibility
and be able to get a great deal of experience quickly. In a large
shop, you're more likely to get a lot of pressure and little
respect. However, a large shop might be better able to offer you
more training and a lot of opportunities for advancement.

A highly experienced locksmith can teach you a lot, but you
probably will have to work a long time to learn much.
Experienced locksmiths place a lot of value on their knowledge
and rarely share much of it with new employees. Often, experienced locksmiths worry that if a new employee learns a lot
quickly, that employee might quit and go to a competitor's shop
or start a competing company.

Some shop owners know little about locksmithing. Working
for them can be difficult because their tools, supplies, and locksmithing methods are outdated. You won't gain much useful
experience working for such a person; instead, you'll probably
pick up a lot of bad habits.

Before attending the National School of Locksmithing and
Alarms in New York City, I worked as an apprentice with a local
locksmith. But I wasn't learning a lot; so I decided to go to school.
I learned a lot at school, and I had my pick of where I wanted to
work. I decided to work with a certified master locksmith who
was a prolific writer. Hardly a month went by without seeing his
technical articles in the trade journals. We often bumped heads,
but working with him was a great learning experience.

Locating Prospective Employers

You can find prospective employers by looking through the
Yellow Pages of telephone directories of the cities in which you
want to work. Your local library probably has lots of out-of-
town directories. Also, look in locksmithing trade journals, and
check out locksmithing web sites. You might also want to place
"job wanted" ads in locksmithing trade journals.

When creating an ad, briefly state the cities in which you would like to work, your most significant qualifications (bondable, good driving record, have own tools, and so forth), and your name, mailing address, and telephone number. Blind advertisements—those that don't include your name, address, or any direct contact information—won't get many responses. Few locksmith-shop owners are that desperate for help.

Getting Prospective Employers to Meet You

It's usually better for you to set up a meeting with a prospective employer than just to walk into the shop unscheduled. You might come in when the owner is busy, which could give the person a negative impression of you. Also, it might be hard to interest the shop owner in hiring you without any prior knowledge of you.

One way to prompt a prospective employer to meet with you is to send a résumé and a cover letter. It isn't always necessary to have a résumé to get a locksmithing job, but it can be a highly effective selling tool. A résumé enables you to project a good image of yourself to a prospective employer, and it sets the stage for your interview.

The résumé is an informative document designed to help an employer know you better. When writing it, view your résumé as a document designed to promote you. Don't include everything there is to know about yourself. Only include information that can help an employer decide to hire you. Don't include, for example, information about being fired from a previous job. It's best to wait until you're face-to-face with a person before you try to explain past problems.

The résumé should be neatly typed in black ink on at least 20-pound bond, 8-1/2 by 11-inch white paper. Figures 19.1 and 19.2 show sample résumés. If you can't type, have someone else type it. Don't use graphics or fancy typestyles (remember, you're seeking work as a locksmith, not as a graphic artist). A résumé should be kept to one page. Few prospective employers like wading through long résumés. There is no perfect structure for a résumé, so organize it in a way that allows the employer to quickly see reasons for hiring you. List your strongest selling points first. If your education is your strongest selling point, for example, include that information first. If your work history is your strongest selling point, include that first.

John A. Smith, RPL, CJS
123 Any Street
Any Town, Any State 01234
Phone: (012) 345-6789

Job Objective
Entry level position as a locksmith in a progressive shop.

Special Capabilities
- Can quickly open locked automobiles
- Can use key cutting and code machines
- Can install, rekey, and service many types of locks
- Can pick locks and impression keys
- Can install and service emergency exit door hardware
- Can operate cash registers

Certifications
Registered Professional Locksmith (RPL)
Certified Journeyman Safecracker (CJS)

Special Qualities
Dependable (missed work only 2 days over the past 5 years), good driver (no accidents), bondable (no arrests), good health, fast learner, and self-starter.

Education/Training
BA degree in Business Administration from ABC College in Quincy, CA. Self-study with *Locksmithing,* by Bill Phillips; and the *Complete Book of Locks & Locksmithing, 6th Edition,* by Bill Phillips.

Memberships
Member of the International Association of Home Safety & Security Professionals; Member of the National Safeman's Organization; Member of the National Locksmith Automobile Association

Work Highlights
OFFICE ASSOCIATE
SEARS IN QUINCY, CA FROM JANUARY 2004 TO PRESENT Man the switchboard; handle complaints; accept money; balance money taken from the safe, registers, and other areas of the store; train other associates; and work with all of the office equipment.

ASSISTANT MANAGER
MCDONALD'S IN QUINCY, CA FROM JUNE 2003 TO JANUARY 2004.

References Available Upon Request

Figure 19.1   If you have a good, relevant work history, emphasize it.

John A. Smith, RPL
123 Any Street
Any Town, Any State 01234
Phone: (012) 345-6789

Job Objective
Seeking position as a locksmith in a locksmithing shop.

Work Experience
April 2005–Present, Manager of ABC Locksmith Shop in Quincy, CA. Supervisor of 6 locksmiths. Duties include: training apprentices; designing and maintaining masterkey systems; installing and servicing emergency exit door devices, electromagnetic locks, electric locks, electric strikes, and a variety of high-security locks; and manipulating, drilling, and recombinating safes.

January 2000–April 2005, Locksmith at DBF Locksmith Shop in Brookdale, CO. Duties included: opening locked automobiles; rekeying, master keying, impressioning, and servicing a variety of basic and high-security locks; manipulating, drilling, and recombinating safes; installing and servicing emergency exit door devices and other door hardware; servicing foreign and domestic automobile locks; installing a wide variety of electric and electronic security devices; operating the cash registers.

December 1997–January 2000, Key Cutter/Apprentice Locksmith at The Key Shop in Brookdale, CO. Cutting keys and assisting three locksmiths. Worked exclusively inside the shop during the first year, but did outside work about half the time thereafter.

June 1995–December 1997, Key Cutter/Salesperson at Building Supply Hardware in Brookdale, CO.

Licenses/Certificates
Have a current California Locksmith Permit
Certified Master Safecracker (CMS)
Member of the International Association of Home Safety and Security Professionals (RPL)
Member of the National Safeman's Organization

References Available Upon Request

Figure 19.2   If you have little experience or formal training, don't emphasize those things on your résumé.

123 Any Street
Any Town, Any State 01234
Phone: (012) 345-6789
Today's Date

Ms. Patricia L. Bruce, owner
The Lock & Key Shop
321 Another Street
Another Town, Another State 43210

Dear Ms. Bruce:

I can perform most basic locksmithing tasks, and especially enjoy doing foreign and domestic automobile work. Please look over my enclosed resume. I'd like to meet with you to see if we might be able to work together.

Please call me at your earliest convenience to let me know when I can come by your shop.

Sincerely yours,
John A. Smith

Figure 19.3 A brief cover letter should be sent with a résumé.

Include a short cover letter with your résumé. The cover letter should be directed to an owner, manager, or supervisor by name and should prompt the person to review your résumé. If you don't know the person's name, call the shop and get it. Don't send a "To Whom It May Concern" cover letter. A sample cover letter is shown in Figure 19.3.

After reviewing your résumé, a prospective employer probably will either call you on the telephone or write you a letter. If you're not contacted within two weeks after mailing your résumé, call the person you addressed your letter to and ask if your letter was received. If it was received, request a meeting. You might be told no job openings are available. In that case, ask if you could still arrange a meeting, so you two can get to know each other.

Preparing to Meet a Prospective Employer

Before going to any meeting, put together a package of documents to leave with the prospective employer. Make photocopies of any relevant certifications, association membership certificates, bond certificate, locksmithing license, and the like. If the prospective employer wants you to complete a job application, attach your documents to it. If they don't ask you to complete a job application, bring your documents to the meeting.

Before the meeting, learn all you can about the prospective employer and their shop. Stop by the shop to find out what products the company sells and what special services it offers. Notice the brands of locks and safes the company sells, and then become familiar with them.

Call the local Better Business Bureau and the Chamber of Commerce to find out how long the business has been established and how many consumer complaints have been filed against it. Contact local, state, and national locksmithing trade associations to find out if the owner belongs to any of them.

Also, before the meeting, read the current issues of major locksmithing trade journals. Study them to find out about major news related to the trade. Then, be prepared to speak confidently about such matters. This will impress the prospective employer.

Before going to the meeting, honestly assess your strengths and weaknesses. Consider which locksmithing tasks you're able to perform well. Also consider how well you can perform other tasks a prospective employer might want you to do, such as selling products, working a cash register, and so on. Then, write a list of all your strengths and a list of the last four places you've worked. Include dates, salaries, and the names of supervisors.

At this point, prepare explanations for any negative questions the prospective employer might ask you during the interview. If there *is* a gap of more than three months between your jobs, for example, decide how to explain that gap. If you were fired from one of your jobs or if you quit one, determine the best way to explain it to the prospective employer. Your explanations should put you in the best light possible.

However, never blame other people, such as former supervisors, for problems in past jobs. This will make you seem like a crybaby or a back stabber. If you don't have a good explanation for being firing, it might be in your best interest simply not to mention that job unless the employer is likely to find out about it anyway. Don't lie about problems you've had because if the prospective employer learns the truth, they probably won't hire you and might tell other locksmiths about you. The locksmithing industry is fairly small, and word gets around quickly.

Meeting a Prospective Employer

On the day of the meeting, be sure you're well rested, have a positive attitude, and are fully prepared. A positive attitude requires seeing yourself as a valuable commodity. Project this image to the person you want to work for. Do not tell a prospective employer that you "really need a job," even if it's true. Few people will hire you simply because you need a job. If you seem too needy, people might think you're incompetent.

Don't go to a meeting with the feeling that the only reason you're there is so a prospective employer can decide if they want to hire you. You both have something to offer that the other needs. You have your time, personality, knowledge, and skills to offer—all of which can help the prospective employer to make more money. The employer, in turn, has knowledge, experience, and money to offer you. The purpose of the meeting is to allow both of you to get to know each other better and to decide if you want to work together. It also might lead to an employment agreement.

When you go to the interview, wear clean clothes and be well groomed. Arrive a few minutes early. When you shake hands with the prospective employer, use a firm (but not tight) grip, look into their eyes, and smile. The person probably will give you a tour of the shop. This can help you to assess whether this is a place where you want to work. Allow the prospective employer to guide the meeting. If they offer you a cup of coffee, decline it unless the prospective employer also has one.

Don't ask if you may smoke and don't smoke if you're invited to. In the meeting, sit in a relaxed position, and look directly into the person's eyes. If you have trouble maintaining eye contact, look at their nose, but try not to look away. Constantly looking away makes you appear insecure or dishonest. Listen intently while the prospective employer is talking, and don't interrupt. Smile a few times during the meeting. Be sure you understand a question before attempting to answer it. Whenever you have the opportunity, emphasize your strengths and what you have to offer, but don't be boastful when doing so.

If you're asked a question you feel uncomfortable answering, keep your body in a relaxed position, continue looking into the person's eyes, and answer it in the way you had planned to answer it. Don't immediately begin talking about another topic after answering the tough question. If you do, this could seem as though you're trying to hide something. Instead, pause after answering the question and smile at the person. They will then either ask you a follow-up question or move on to another topic. Don't give an audible sigh of relief when you move to a new topic.

If asked about your salary requirements, don't give a figure. Instead, say something like, "I don't have a salary in mind, but I'm sure we can agree on one if we decide to work together." Salary is usually discussed at a second meeting. However, if the prospective employer insists on discussing salary during the first meeting, let them state a figure first. Ask what they believe is a fair salary. When discussing salary, always speak about "fairness to both of us."

Making an Employment Agreement

The prospective employer's first salary offer almost always will be less than the amount the company is willing to pay. You might be told the employer doesn't know about your ability to work, so they will start you off at a low salary that can be reviewed later.

This might sound good, but the salary you start out with has a lot to do with how much you earn from the company later. You want to begin working for the highest salary possible.

20

The Safe and Secure Home

This chapter discusses how you can put everything together to make a home as safe as the apartment or homeowner wants it to be. It's important to understand that no single security plan is best for everyone. Each home and neighborhood has unique strengths and vulnerabilities, and each household has different needs and limitations.

The important limitation most of us face is money. If money were no object, it would be easy to lay out a great security plan for you, with redundant security, such as is found at casinos. Those purchases would make any home more secure but, in all likelihood, they would be overkill.

With proper planning, you can make a home more secure and stay within your customer's budget, without inconveniencing your customer. Proper planning is based on the following considerations:

1. How much money is your customer willing to spend?

2. How much risk is acceptable to your customer?

3. How much inconvenience is acceptable to your customer?

Before you can create the best plan for your customer, you need to conduct a safety and security survey (or "vulnerability analysis"). The survey requires you to walk around the outside of the home and through every room. You should make note of all potential problems.

Surveying a House

The purposes of a safety and security survey are:

1. To help you identify potential problems.
2. To assess how likely and how critical each risk is.
3. To determine cost-effective ways to either eliminate the risks or bring them to an acceptable level.

The survey enables you to take precise, integrated security and safety measures.

A thorough survey involves not only inspecting the inside and outside of the home, but also examining all the safety and security equipment, and then reviewing the safety and security procedures used by all members of the household. The actions people take (or fail to take) are just as important as the equipment you may buy. What good are high-security deadbolts, for instance, if residents often leave the doors unlocked?

As you conduct your survey, keep the information in the preceding chapters in mind. You'll notice many potential safety and security risks (every home has some). Some of the risks are simple to reduce or eliminate immediately. For others, you need to compare the risk to the cost of properly dealing with it. You don't have a mathematical formula to fall back on. You need to make subjective decisions, based on what you know about the household and neighborhood, and guided by the information throughout this book.

When surveying the house, starting outside is best. Walk around the house and stand at the vantage points passersby are likely to have. Many burglars target a home because it's especially noticeable while driving or walking past it. When you look at the home from the street, note any feature that may make someone think no one is home, that a lot of valuables could be inside, or that the home might be easy to break into.

Remember, burglars prefer to work in secrecy. They like heavy shrubbery or large trees that block or crowd an entrance, and they like homes that aren't well-lighted at night. Other things that may attract burglars' attention include expensive items that can be seen through windows, a ladder near the home, and notes tacked on the doors.

As you walk around the home, note anything that might help discourage burglars. Can your "Beware of Dog" sign or your fake security system sticker be seen in the window? Walk to each entrance and consider what burglars might like and dislike about it. Is the entrance well-lighted? Can neighbors see someone who's at the entrance? Is there a video camera pointing at the entrance? Does the window or door appear easy to break into?

After surveying the outside of the home, go inside and carefully examine each exterior door, window, and other opening. Consider whether each one is secure, but allows residents to get out quickly. Check for the presence of fire safety devices. Does the family have enough smoke detectors and fire extinguishers in working order? Are they in the best locations?

Take an honest look at the safety and security measures the family takes. What habits or practices might need to be changed?

Surveying an Apartment

In many ways, surveying an apartment is like surveying a house. The difference is you must be concerned not only about the actions of one household, but also about those of the landlord, the apartment managers, and the other tenants. The less security-conscious others around your customers are, the more at-risk your customer is. No matter how much you do to avoid causing a fire, for example, a careless neighbor may cause one. If your customer's neighbors don't care about crime prevention, they may attract burglars to your neighborhood. But, if your customer has taken the security measures in this book, your customer's apartment will be less attractive to burglars.

As you walk around the outside of an apartment, notice everything that would-be burglars might notice. Will they see tenants' "crime watch" signs? Will they see that all the apartments have door viewers and deadbolt locks? Burglars hate a lot of door viewers because they never know when someone might be watching them.

After surveying the outside, walk through the apartment and look at each door, window, and other opening. If you notice major safety or security problems, point them out to your customer. You might also want to suggest little things that the customer or landlord can do to make the apartment more secure.

High-Rise Apartment Security

High-rise apartments have special security concerns that don't apply to apartments with fewer floors. In a high-rise, more people have keys to the building, which means more people can carelessly allow unauthorized persons to enter.

The physical structure of a high-rise often provides many places for criminals to lie in wait for victims or to break into apartments unnoticed. Many high-rise buildings aren't designed to allow people to escape quickly during a fire.

The safest apartments have only one entrance for tenants to use, and that entrance is guarded 24 hours a day by a doorman. An apartment that doesn't have a doorman should have a video intercom system outside the building. Video intercoms are better than audio intercoms because they let you see and hear who's at the door before buzzing the person in.

It's also important to consider the building's fire exits. Every fire-exit door leading directly to apartments should stay locked from the fire-exit side. In other words, a person who's in a fire-exit stairwell should have to walk out of the building and reenter through the main entrance to get onto any floor. The fire exits should be only for getting out—not for getting in.

All the fire-exit doors on the main floor should be connected to an alarm that sounds if someone opens one of the doors and that can be heard by the doorman (or by a maintenance person).

Exterior fire escapes—the metal stairs on the outside of a building—present another problem. Many of them have ladders that can be reached from the ground. Burglars can use a stick to disengage the hook that holds the ladder, and then make the ladder slide down. They can then climb the ladder to get into any apartment accessible via the fire escape.

Apartments that can be accessed from an exterior fire escape should have grates inside that prevent anyone from climbing through the windows. All exterior doors leading directly to a fire escape should be strong and equipped with good locks.

Elevators should be programmed, so they always stop at the ground floor before descending to the basement and before going to the upper floors. Too often, an elevator passenger is overcome by an attacker and taken directly to the basement or to another floor to be assaulted. If the elevator always stops at the ground floor, a captive has a better chance of getting out or of being seen.

Elevators should have corner-mounted mirrors that let you see who's on the elevator before you get on. The mirrors should be positioned so there is no hiding spot in the elevators. If you see someone who makes you feel uncomfortable, you can avoid getting onto the elevator.

As with any type of apartment, you should consider how security-conscious the neighbors are. In a high-rise, security depends largely on whether the building's management stresses the importance of security to the tenants and reprimands flagrant violations.

One problem in many high-rises is that burglars are let into the building by a tenant who doesn't bother checking the identity of people before buzzing them in. Some burglars get into buildings by *piggybacking*—running in behind a tenant who has a key to the door. Many tenants don't like to question a stranger who piggybacks for fear of being attacked.

Home Safety and Security Checklist

As you conduct your survey, note each potential problem that is of concern to you.

Home Exterior

_____ Shrubbery. (Shouldn't be high enough for a burglar to hide behind—or too near windows or doors.)

_____ Trees. (Shouldn't be positioned so a burglar can use them to climb into a window.)

_____ House numbers. (Should be clearly visible from the street.)

_____ Entrance visibility. (Should allow all entrances to be seen clearly from the street or other public area.)

_____ Lighting near garage and other parking areas.

_____ Ladders. (Shouldn't be in the yard or in clear view.)

_____ "Alarm System" or "Surveillance System" stickers. (Shouldn't identify the type of system that's installed.)

_____ Mailbox. (Should be locked or otherwise adequately secured, and should show no name or only a first initial and last name.)

_____ Windows. (Should be secured against being forced open, but should allow for easy emergency escape.)

_____ Window air conditioners. (Should be bolted down or otherwise protected from removal.)

_____ Fire escapes. (Should allow for easy emergency escape but not allow unauthorized entry.)

Exterior Doors and Locks

(Included here are doors connecting a garage to the home.)

_____ Door material. (Should be solid hardwood, fiberglass, PVC plastic, or metal.)

_____ Door frames. (Should allow doors to fit snugly.)

_____ Door glazing. (Shouldn't allow someone to gain entry by breaking it and reaching in.)

_____ Door viewer (without glazing). (Should have a wide-angle door viewer or other device to see visitors.)

_____ Hinges. (Should be either on inside of door or protected from outside removal.)

_____ Stop molding. (Should be one-piece or protected from removal.)

_____ Deadbolts. (Should be single-cylinder with free-spinning cylinder guard and a bolt with a 1-inch throw and a hardened insert.)

_____ Strike plates. (Should be securely fastened.)

_____ Door openings (mail slots, pet entrances, and other access areas). (Shouldn't allow a person to gain entry by reaching through them.)

_____ Sliding glass doors. (Should have a movable panel mounted on interior side, and a bar or other obstruction in the track.)

Inside the Home

_____ Fire extinguishers. (Should be in working order and mounted in easily accessible locations.)

_____ Smoke detectors. (Should be in working order and installed on every level of the home.)

_____ Rope ladders. (Should be easily accessible to bedrooms located above the ground floor.)

_____ Flashlights. (Should be in working order and readily accessible.)

_____ First-aid kit. (Should contain fresh bandages, wound-dressing and burn ointments, aspirin, and plastic gloves.)

_____ Telephones. (Should be programmed to dial the police and fire departments quickly or their phone numbers should be posted nearby.)

_____ Burglar alarm. (Should be in good working order and adequately protected from vandalism, and should have adequate backup power.)

_____ Safes. (Should be installed so they can't be seen by visitors.)

People in the Home

_____ Doors (locking). (Should be locked by all residents every time they leave the home—even if they plan to be gone for only a few minutes.)

_____ Doors (opening). (Should not be opened by any resident unless the person seeking entry has been satisfactorily identified.)

_____ Fire escape plan. (Should be familiar to all residents as a result of practicing how to react during a fire.)

_____ Confidentiality. (Should be maintained by all residents regarding locations of safes, burglar alarms, and other security devices.)

_____ Money and valuables. (Should be stored in the home only in small amounts and only if they have low resale or "fencing" value.)

_____ Drapes and curtains. (Should be routinely closed each night.)

_____ Garage doors. (Should be kept closed and locked.)

21

Security Consulting

As a locksmith, you are in a position to sell your consulting services. When I started consulting, I quickly began getting all kinds of work. I get new and exciting offers all the time to consult for large corporations, law firms, web sites, and periodicals.

If you're flexible, you can sell your consulting services in many ways. One major lock manufacturer hired me to travel around the country doing home security surveys in low-income neighborhoods and installing new locks. *Consumers Digest* hired me to conduct tests on door locks, safes, and car alarms, as well as to make recommendations for best buys. I've also written advertorials for companies. A law firm hired me to consult about lock-opening techniques.

This chapter tells how to begin selling your consulting services. One option is to write articles for trade journals, such as the *National Locksmith*. That experience can help you become a better writer, and it can help you approach general consumer publications, which pay much more than trade journals. You'll have a better chance of writing for general consumer publications if you have a few clips from trade journals. My first article was for *Locksmith Ledger* and I wrote it while I was still in locksmithing school. The article helped me get a job right out of school.

I enjoyed the experience of seeing my name in print, and started writing for trade journals, including *Security Dealer*, the *Guild Report*, *Keynotes*, and many others. After about a year of writing for the trade, I was approached by a book publisher about revising a locksmithing book. After revising that book, I started writing for the general consumer. Since then, I've written hundreds of articles and dozens of books, for the trade and general consumers.

Writing articles and books can help build your reputation as a locksmith, which is necessary if you want to do security consulting. I've gotten jobs from people who read one of my books or articles.

To begin writing for trade magazines, first come up with a good article idea, and write or e-mail the magazine's editor. If you're doing a locksmithing job, take a lot of photographs from start to finish. Good photographs impress an editor.

After writing articles, you may want to write a book. Again, the more clips you have, the more it can impress an editor that you're the one to write the book. When you're ready, write a query letter to an acquisition editor at a publishing house that publishes security-related books, such as McGraw-Hill. Writing books isn't for everyone, however. Some people have great ideas, but they have a hard time putting those ideas down on paper. If that's the case, you may want to cowrite an article or book. Find someone to work with who has already written a security book.

Consulting Fees

How much should you charge for consulting? As a rule, you don't want to charge more than or less than everyone else. The lowest rate I know of is $75 an hour. The highest is $300 an hour. I charge $200 plus expenses per hour for consulting, paperwork, and research. When I'm traveling, I charge $100 an hour, plus expenses. Expenses include all travel expenses and meals. I usually have the client pay for my airline ticket in advance. I also require a retainer upfront, such as $1,000. I've never had a problem getting clients to accept my fees. A sample invoice is shown in Figure 21.1.

Confidentiality

When you're working as a security consultant, you may or may not be asked to sign a confidentiality agreement. Even if you're not asked to sign one, you are responsible for keeping confidential sensitive information that you obtain as a result of your work. I never share names of clients or what work I've

Bill Phillips
Box 2044
Erie, PA 16512-2044
Phone: 814-453-7154
Email: LocksmithWriter@aol.com
March 4, 2007

Mr. John Smith
Smith & Jones P.C.
Any Address
Any City, Any State 00000

Re: Invoice for security consulting services rendered

Dear John:

Here is my invoice for services rendered on January 24 through January 27, 2007.

1 hour of research time @ $200/hour.	$200.00
Taxi to and from Detroit Airport.	$57.00
Taxi to and from Erie Airport.	$30.00
Food. Breakfast and dinner.	$15.89
Travel time (to Detroit) 8:30 am to 2:00 pm 5.5 hours @ $100/hour.	$550.00
Meeting time. 2:00 pm to 4:00 pm. 2 hours @ $200/hour.	$400.00
Travel time (to Erie) 4:00–9:30 pm. 5.5 hours @ $100/hour.	$550.00
Total	$1802.89
Less Retainer	$1000.00
Total Now Due	**$802.89**

Sincerely,

Bill Phillips

Figure 21.1 An invoice for security consulting.

done for them, unless my client says it's fine to share the information.

Another thing to remember, especially when working for a law firm, is not to conduct any tests or write any reports unless the firm requests it.

When entering into a confidentiality agreement, make sure it's not so broad that it prevents you from writing other locksmithing or security-related articles and books. Keep such an agreement as broad as possible. Limit confidentiality to "sensitive or proprietary information that was learned during the course of the security consulting work."

Getting Started in Consulting Work

If you want to be a security consultant, you first need to determine what kind of clients you want to target. Then, have some stationery made that says you're a security consultant. Having a graphic designer do this for you is worth the expense. If you can't afford to have professional stationery designed, you can use a software program. At a minimum, you need letterhead and a business card. A trifold folder that gives more information about you and your services can also be helpful.

Before mailing anything, find out the name and address of the person who can make the decision to use your services. Then send your information to that person by name. Send them a short letter on your letterhead, a business card, and a brochure (if you have one).

BHMA
Finish Codes

600-Primed For Painting

601-Bright Black Japanned

603-Zinc Plated

604-Zinc Plated and Dichromate Sealed

605-Bright Brass, Clear Coated

606-Satin Brass, Clear Coated

609-Satin Brass, Blackened, Satin Relieved, Clear Coated

610-Satin Brass, Blackened, Bright Relieved, Clear Coated

611-Bright Bronze, Clear Coated

612-Satin Bronze, Clear Coated

616-Satin Bronze, Blackened, Satin Relieved, Clear Coated

618-Bright Nickel Plated, Clear Coated

619-Satin Nickel, Clear Coated

620-Satin Nickel Plated, Blackened, Satin Relieved, Clear Coated

621-Nickel Plated, Blackened, Relieved, Clear Coated

625-Bright Chromium Plated

626-Satin Chromium Plated

629-Bright Stainless Steel

630-Satin Stainless Steel

632-Bright Brass Plated, Clear Coated

633-Satin Brass Plated, Clear Coated

636-Satin Brass Plated, Blackened, Bright Relieved, Clear Coated

637-Bright Bronze Plated, Clear Coated

638-Satin Brass Plated, Blackened, Bright Relieved, Clear Coated

639-Satin Bronze Plated, Clear Coated

643-Satin Bronze Plated, Blackened, Bright Relieved, Clear Coated

645-Bright Nickel Plated, Clear Coated

646-Satin Nickel Plated, Clear Coated

647-Satin Nickel Plated, Blackened, Bright Relieved, Clear Coated

648-Nickel Plated, Blackened, Relieved, Clear Coated

651-Bright Chromium Plated

652-Satin Chromium Plated

Standard Cylinders

Arrow

Shoulder to First Cut: .265"
Center to Center: .155"
Macs: 7

	Root Depth	Bottom Pins	Master Pins
#0	.312	.178	
#1	.298	.192	
#2	.284	.206	.028
#3	.270	.220	
#4	.256	.234	.056
#5	.242	.248	
#6	.228	.262	.084
#7	.214	.276	
#8	.200	.290	.112
#9	.186	.304	

Corbin

	Z Bow	X Bow
Shoulder to First Cut:	.250"	.197"

Center to Center: .156"
Macs: 8

	Root Depth		Bottom Pins		Master Pins
	Z Bow	X Bow	Z Bow	X Bow	
#1	.343	.333	.160	.171	
#2	.329	.319	.174	.185	.028
#3	.315	.305	.189	.198	.042
#4	.301	.291	.203	.212	.056
#5	.287	.277	.218	.226	.069
#6	.273	.263	.231	.241	.084
#7	.259	.249	.246	.256	.099
#8	.245	.235	.259	.269	.112
#9	.231	.221	.273	.284	.127
#10	.217	.207	.287	.297	

Dexter

Shoulder to First Cut: .216"
Center to Center: .155"
Macs: 7

	Root Depth	Bottom Pins	Master Pins
#0	.320	.165	
#1	.305	.180	
#2	.290	.195	.030
#3	.275	.210	
#4	.260	.225	.060
#5	.245	.240	
#6	.230	.255	.090
#7	.215	.270	
#8	.200	.285	.120
#9	.185	.300	

Emhart

Shoulder to First Cut: .250"
Center to Center: .156"
Macs: Dependent on Adjacent Angles

	Root Depth	Bottom Pins	Master Pins	Drivers
#1	.193			
#2	.305	.242	.097	.158
#3	.277	.270	.125	
#4	.249	.298	.153	
#5	.221	.326		
#6	.193	.354		

Emhart High-Security Keying Restrictions

1. No #1 cuts allowed—cut will be too shallow to maintain angle

   - Even though these cylinders are compatible
   - With system 70, progression is limited to 2–6
   - Mixing these-high security cylinders with conventional system 70 cylinders must be carefully planned from inception—please refer to the Corbin/Russwin Cylinder Manual for more information.

2. There is no #1 master pin—therefore change key cut must be at least 2 steps from master cut.

3. Macs is dependent on adjacent angles—when adjacent angles are the same, Macs is 4; when adjacent angles are different, Macs is 3.

4. When the bottom and master pins total 2, 3, or 4, the driver is a #1; when the bottom and master pins total 5 or 6, the driver is a #2.

Falcon

Shoulder to First Cut: .237"
Center to Center: .156"
Macs: 7

	Root Depth	Bottom Pins	Master Pins
#0	.315	.168	
#1	.297	.186	
#2	.279	.204	.036
#3	.261	.222	
#4	.243	.240	.072
#5	.225	.258	
#6	.207	.276	.108
#7	.189	.294	
#8	.171	.312	.144
#9	.153	.330	

ILCO

Shoulder to First Cut: .277"
Center to Center: .156"
Macs: 7

	Root Depth	Bottom Pins	Master Pins
#0	.320	.180	
#1	.302	.198	
#2	.284	.216	.036
#3	.266	.234	
#4	.248	.252	.072
#5	.230	.270	
#6	.212	.288	.108
#7	.194	.306	
#8	.176	.324	.144
#9	.158	.342	

Kwikset

Shoulder to First Cut: .247"
Center to Center: .150"

Titan—Shoulder to First Cut: .097"
Center to Center: .150"
Macs: 4

	Root Depth	Bottom Pins	Master Pins
#1	.329	.172	.023
#2	.306	.195	.046
#3	.283	.218	.069
#4	.260	.241	.092
#5	.237	.264	.115
#6	.214	.287	*
#7	.191	.310	

Kwikset does not make a #6 master wafer.

Lockwood

Shoulder to First Cut: .277"
Center to Center: .156"
Macs: 7

	Root Depth	Bottom Pins	Master Pins
#0	.320	.150	
#1	.305	.165	
#2	.290	.180	.030
#3	.275	.195	.045
#4	.260	.210	.060
#5	.245	.225	.075
#6	.230	.240	.090
#7	.215	.355	.105
#8	.200	.370	.120
#9	.185	.385	.135

Medeco

Shoulder to First Cut: .244"
Center to Center: .170"
Macs: 4

	Root Depth	Bottom Pins	Master Pins	Drivers
#1	.266	.236	.030	.270
#2	.236	.266	.060	.240
#3	.206	.296	.090	.210
#4	.176	.326	.120	.180
#5	.146	.356	.150	.150
#6	.116	.386	.120	

Medeco does not allow change keys with a #6 cut next to the shoulder.

Medeco Biaxial

Shoulder to First Cut: Fore: .213"
　　　　　　　　　　　Aft: .275"
Center to Center: Aft—Fore: .108"
　　　　　　　　　Fore—Fore: .170"
　　　　　　　　　Aft—Aft: .170"
　　　　　　　　　Fore—Aft: .232"
Macs: Aft—Fore: 2
　　　　Fore—Fore: 3
　　　　Aft—Aft: 3
　　　　Fore—Aft: 4

	Root Depth	Bottom Pins	Master Pins	Drivers
#1	.264	.239	.025	.270
#2	.239	.264	.050	.240
#3	.214	.289	.075	.210
#4	.189	.314	.100	.180
#5	.164	.339	.125	.150
#6	.139	.364		

Medeco Keymark

Shoulder to First Cut: .195"
Tip to First Cut: .090"
Cut to Cut: .150"
Macs: 9

	Root Depth	Bottom Pins	Master Pins
#0	.1385	.110	—
#1	.1260	.122	—
#2	.1135	.135	.025
#3	.1010	.147	.037
#4	.0885	.160	.050
#5	.0760	.172	.062
#6	.0635	.185	.075
#7	.0510	.197*	.087
#8	.0385	.210*	.100
#9	.0260	.222*	.112

Spool Pins

Russwin

Shoulder to First Cut: .250"
Center to Center: .156"
Macs: 7

	Root Depth	Bottom Pins	Master Pins
#0	.341	.160	
#1	.326	.175	
#2	.311	.189	.030
#3	.296	.203	.045
#4	.281	.220	.060
#5	.266	.234	.075
#6	.251	.248	.090
#7	.236	.263	.105
#8	.221	.279	.120
#9	.206	.294	.135

Sargent

Shoulder to First Cut: .216"
Center to Center: .156"
Macs: 7

	Root Depth	Bottom Pins	Master Pins
#1	.328	.170	
#2	.308	.190	.040
#3	.288	.210	.060
#4	.268	.230	.080
#5	.248	.250	.100
#6	.228	.270	.120
#7	.208	.290	.140
#8	.188	.310	.160
#9	.168	.330	.180
#10	.148	.350	

Schlage

Shoulder to First Cut: .231"
Center to Center: .156"
Macs: 7

	Root Depth	Bottom Pins	Master Pins
#0	.335	.165	
#1	.320	.180	
#2	.305	.195	.030
#3	.290	.210	
#4	.275	.225	.060
#5	.260	.240	
#6	.245	.255	.090
#7	.230	.270	
#8	.215	.285	.120
#9	.200	.300	

Segal

Shoulder to First Cut: .262"
Center to Center: .156"
Macs: 5

	Root Depth	Bottom Pins	Master Pins
#0	.315	.166	
#1	.295	.186	.020
#2	.275	.206	.040
#3	.255	.226	.060
#4	.235	.246	.080
#5	.215	.266	.100
#6	.195	.286	*

No #6 master wafer

System 70

Shoulder to First Cut: .250"
Center to Center: .156"
Macs: 4

	Root Depth	Bottom Pins	Master Pins
#1	.339	.160	.028
#2	.311	.189	.056
#3	.283	.217	.084
#4	.255	.245	.112
#5	.227	.273	.140
#6	.199	.301	

Weiser

Shoulder to First Cut: .237"
Center to Center: .156"
Macs: 7

	Root Depth	Bottom Pins	Master Pins
#0	.315	.168	
#1	.297	.186	
#2	.279	.204	.036
#3	.261	.222	
#4	.243	.240	.072
#5	.225	.258	
#6	.207	.276	.108
#7	.189	.294	
#8	.171	.312	.144
#9	.153	.330	

Weslock

Shoulder to First Cut: .220"
Center to Center: .156"
Macs: 7

	Root Depth	Bottom Pins	Master Pins
#0	.330	.156	
#1	.314	.172	
#2	.299	.187	.030
#3	.283	.202	
#4	.268	.219	.060
#5	.252	.234	
#6	.236	.250	.090
#7	.221	.265	
#8	.205	.281	.120
#9	.190	.297	

Yale

Shoulder to First Cut: .206"
Center to Center: .165"
Macs: 7

	Root Depth	Bottom Pins	Master Pins
#0	.320	.182	
#1	.301	.201	
#2	.282	.220	.038
#3	.263	.239	
#4	.244	.258	.076
#5	.225	.277	
#6	.206	.296	.114
#7	.187	.315	
#8	.168	.334	.152
#9	.149	.353	

Interchangeable Core

A2 (Arrow, Falcon, Eagle, Best)

Macs: 9

	Root Depth	Bottom Pins	Master Pins	Control Pins	Drivers
#0	.318	.110	—	—	
#1	.305	.122	—	—	
#2	.293	.135	.025	.025	
#3	.280	.147	.037	.037	
#4	.268	.160	.050	.050	.050
#5	.255	.172	.062	.062	.062
#6	.243	.185	.075	.075	.075
#7	.230	.197	.087	.087	.087
#8	.218	.210	.100	.100	.100
#9	.205	.222	.112	.112	.112
#10	.125	.125			
#11	.137	.137			
#12	.150	.150			
#13	.162	.162			

(continued on next page)

	Root Depth	Bottom Pins	Master Pins	Control Pins	Drivers
#14	.175				
#15	.187				
#16	.200				
#17	.212				
#18	.225				
#19	.237				

If the master cut is even, make the corresponding control cut odd. If the master cut is odd, make the corresponding control cut even.

Drivers: To calculate drivers, add 10 to the control key cut and subtract from 23. This calculation is constant throughout the system.

A3 (Arrow, Falcon, Eagle, Best)

Macs: 6

	Root Depth	Bottom Pins	Master Pins	Control Pins	Drivers
#0	.318	.110			
#1	.300	.128	.018	.018	
#2	.282	.146	.036	.036	
#3	.264	.164	.054	.054	
#4	.246	.182	.072	.072	.072
#5	.228	.200	.090	.090	.090
#6	.210	.218	.108	.108	.108
#7	.126	.126			
#8	.144	.144			
#9	.162	.162			
#10	.180	.180			
#11	.198	.198			
#12	.216				
#13	.234				

Drivers: To calculate drivers, add 7 to the control key cut and subtract from 16. This calculation is constant throughout the system.

Many A3 systems do not use change keys with cuts within one step of corresponding master cut.

A4 (Arrow, Falcon, Eagle, Best)

Macs: 5

	Root Depth	Bottom Pins	Master Pins	Control Pins	Drivers
#0	.318	.110			
#1	.297	.131	.021	.021	
#2	.276	.152	.042	.042	
#3	.255	.173	.063	.063	.063
#4	.234	.195	.084	.084	.084
#5	.213	.216	.105	.105	.105
#6	.126	.126			
#7	.147	.147			
#8	.168	.168			
#9	.189	.189			
#10	.210	.210			
#11	.231	.231			
#12	.252				

Drivers: To calculate drivers, add 6 to the control key cut and subtract from 14. This calculation is constant throughout the system.

Corbin

	Z Bow	X Bow
Shoulder to First Cut:	.250"	.197"

Center to Center: .156"
Macs: 8

	Root Depth		Bottom Pins		Master Pins	Control Pins	Drivers
	Z Bow	X Bow	Z Bow	X Bow			
						−9=.037	
						−8=.051	0=.247
#1	.339	.333	.160	.171		−7=.066	.192
#2	.325	.319	.174	.185	.028	−6=.080	.177
#3	.311	.305	.189	.198	.042	−5=.093	.163
#4	.297	.291	.203	.212	.056	−4=.107	.149

(continued on next page)

	Root Depth		Bottom Pins		Master Pins	Control Pins	Drivers
	Z Bow	X Bow	Z Bow	X Bow			
#5	.283	.277	.218	.226	.069	−3=.120	.135
#6	.269	.263	.231	.241	.084	−2=.135	.120
#7	.255	.249	.246	.256	.099	−1=.149	.107
#8	.241	.235	.259	.269	.112	0=.163	.093
#9	.227	.221	.273	.284	.127	+1=.177	.080
#10	.213	.207	.287	.297		+2=.192	.066
						+3=.205	
						+4=.218	
						+5=.232	
						+6=.247	
						+7=.261	
						+8=.275	
						+9=.289	

Non-control chambers use a #0 driver.

In the control chambers the drivers are the same number as the corresponding control key cut.

Emhart

Shoulder to First Cut: .250"
Center to Center: .156"
Macs: Dependent on Adjacent Angles

	Root Depth	Bottom Pins	Master Pins	Drivers	Control Chambers (Use conventional pins)			
					Bottom Pins	Control Pins	Master Pins	Drivers
#1				.193		−4=.030	.028	
#2	.305	.242	.097	.158	.231	−3=.058	.056	.198
#3	.277	.270	.125		.260	−2=.087	.084	.171
#4	.249	.298	.153		.288	−1=.114	.112	.142
#5	.221	.326			.316	0=.142		.114
#6	.193	.354			.344	+1=.171		.087
						+2=.198		
						+3=.226		
						+4=.253		

Emhart High-Security Keying Restrictions

1. No #1 cuts allowed—cut will be too shallow to maintain angle

   • Even though these cylinders are compatible with system 70, progression is limited to 2–6.

   • Mixing these high-security cylinders with conventional system 70 cylinders must be carefully planned from inception—please refer to the Corbin/Russwin Cylinder Manual for more information.

2. There is no #1 master pin—therefore change key cut must be at least 2 steps from master cut.

3. Macs is dependent on adjacent angles—when adjacent angles are the same, Macs is 4; when adjacent angles are different, Macs is 3.

4. Change key cuts cannot be the same depth as control key cuts in both control chambers, regardless of the angles.

5. When the bottom and master pins total 2, 3, or 4, the driver is a #1; when the bottom and master pins total 5 or 6, the driver is a #2. This does not apply to the control chambers. See #6 below.

6. Control chambers use conventional pins, therefore they use different drivers and build-up pins.

KABA A2

140 Spacing—From Tip—.136"
             From Bow—1.030"
             Cut to Cut—.140
Macs: 8

150 Spacing—From Tip—.086"
             From Bow—1.080"
             Cut to Cut—.150
Macs: 9

	Root Depth	Bottom Pins	Master Pins	Control Pins	Drivers
#0	.318	.110	—	—	
#1	.305	.122	—	—	
#2	.293	.135	.025	.025	
#3	.280	.147	.037	.037	
#4	.268	.160	.050	.050	.050
#5	.255	.172	.062	.062	.062
#6	.243	.185	.075	.075	.075
#7	.230	.197	.087	.087	.087
#8	.218	.210	.100	.100	.100
#9	.205	.222	.112	.112	.112
#10	.125	.125			
#11	.137	.137			
#12	.150	.150			
#13	.162	.162			
#14	.175				
#15	.187				
#16	.200				
#17	.212				
#18	.225				
#19	.237				

Drivers: To calculate drivers, add 10 to the control key cut and subtract from 23. This calculation is constant throughout the system.

Some depths in position #6 (next to Peak) require an asymmetrical cutter: A2—7, 8, and 9
             A4—4 and 5

KABA A4

140 Spacing—From Tip—.136"
                From Bow—1.030"
                Cut to Cut—.140

Macs: 4

150 Spacing—From Tip—.086"
                From Bow—1.080"
                Cut to Cut—.150

Macs: 5

	Root Depth	Bottom Pins	Master Pins	Control Pins	Drivers
#0	.318	.110			
#1	.297	.131	.021	.021	
#2	.276	.152	.042	.042	
#3	.255	.173	.063	.063	.063
#4	.234	.195	.084	.084	.084
#5	.213	.216	.105	.105	.105
#6	.126	.126			
#7	.147	.147			
#8	.168	.168			
#9	.189	.189			
#10	.210	.210			
#11	.231	.231			
#12	.252				

Drivers: To calculate drivers, add 6 to the control key cut and subtract from 14. This calculation is constant throughout the system.

Some depths in position #6 (next to Peak) require an asymmetrical cutter: A2—7, 8, and 9
                A4—4 and 5

Medeco

Shoulder to First Cut: .244"
Center to Center: .170"
Macs: 4

	Root Depth	Bottom Pins	Master Pins	Drivers
#1	.266	.236	.030	.270
#2	.236	.266	.060	.240
#3	.206	.296	.090	.210
#4	.176	.326	.120	.180
#5	.146	.356	.150	.150
#6	.116	.386	.120	

The stack height is .506".

Drivers are the same number as the corresponding change key cut.

Medeco International Core Keying Restrictions

Medeco does not use control pins in the two control chambers. The control key cuts in these chambers must be three steps shallower than the master cuts in these positions. Multi-core does this automatically.

1. A #6 cut cannot be used next to the shoulder.

2. Master key cuts in chambers 3 and 4 must be 4, 5, or 6.

3. Change key cuts in chamber 3 or 4 must be restricted, depending on the master cuts.

Master Cut	Change Key Possibilities
4	2 and 3 or 2 and 6 or 3 and 5 or 5 and 6
5	1 and 3 or 1 and 6 or 3 and 4 or 4 and 6
6	1 and 2 or 2 and 4 or 1 and 5 or 4 and 5

4. Change key cuts cannot be the same as control key cuts in both control positions.

Medeco Biaxial ICore

Shoulder to First Cut: Fore: .213"
                          Aft: .275"
Center to Center: Aft—Fore: .108"
                          Fore—Fore: .170"
                          Aft—Aft: .170"
                          Fore—Aft: .232"
Macs: Aft—Fore: 2
          Fore—Fore: 3
          Aft—Aft: 3
          Fore—Aft: 4

	Root Depth	Bottom Pins	Master Pins	Drivers
#1	.264	.239	.025	.270
#2	.239	.264	.050	.240
#3	.214	.289	.075	.210
#4	.189	.314	.100	.180
#5	.164	.339	.125	.150
#6	.139	.364		

Medeco International Core Keying Restrictions

Medeco does not use control pins in the two control chambers. The control key cuts in these chambers must be three steps shallower than the master cuts in these positions. Multi-core does this automatically.

1. A #6 cut cannot be used next to the shoulder.

2. Master key cuts in chambers 3 and 4 must be 4, 5, or 6.

3. Change key cuts in chamber 3 or 4 must be restricted, depending on the master cuts.

Master Cut	Change Key Possibilities
4	2 and 3 or 2 and 6 or 3 and 5 or 5 and 6
5	1 and 3 or 1 and 6 or 3 and 4 or 4 and 6
6	1 and 2 or 2 and 4 or 1 and 5 or 4 and 5

4. Change key cuts cannot be the same as control key cuts in both control positions.

Medeco Keymark

Shoulder to First Cut: .195"
Tip to First Cut: .090"
Cut to Cut: .150"
Macs: 9

	Root Depth	Bottom Pins	Master Pins	Control Pins	Drivers
#0	.1385	.110	—	—	
#1	.1260	.122	—	—	
#2	.1135	.135	.025	.025	
#3	.1010	.147	.037	.037	
#4	.0885	.160	.050	.050	.050
#5	.0760	.172	.062	.062	.062
#6	.0635	.185	.075	.075	.075
#7	.0510	.197*	.087	.087	.087
#8	.0385	.210*	.100	.100	.100
#9	.0260	.222*	.112	.112	.112
#10	.125	.125			
#11	.137*	.137			
#12	.150	.150			
#13	.162*	.162			
#14	.175				
#15	.187*				
#16	.200				
#17	.212*				
#18	.225				
#19	.237				

Spool Pins

If the master cut is even, make the corresponding control cut odd.

If the master cut is odd, make the corresponding control cut even.

Russwin

Shoulder to First Cut: .250"
Center to Center: .156"
Macs: 7

	Root Depth	Bottom Pins	Master Pins	Control Pins	Drivers
#0	.341	.160		−9=.028	.192
#1	.326	.175		−8=.042	.177
#2	.311	.189	.030	−7=.058	.163
#3	.296	.203	.045	−6=.072	.149
#4	.281	.220	.060	−5=.087	.133
#5	.266	.234	.075	−4=.103	.118
#6	.251	.248	.090	−3=.118	.103
#7	.236	.263	.105	−2=.133	.087
#8	.221	.279	.120	−1=.149	.072
#9	.206	.294	.135	0=.163	.058
				+1=.177	
				+2=.192	
				+3=.208	
				+4=.222	
				+5=.238	
				+6=.253	
				+7=.268	
				+8=.282	
				+9=.298	

For non-control chambers use .247".

Sargent

Shoulder to First Cut: .216"
Center to Center: .156"
Macs: 7

	Root Depth	Bottom Pins	Master Pins	Control Pins	Drivers
#1	.328	.170			
#2	.308	.190	.040	.040	.040
#3	.288	.210	.060	.060	.060
#4	.268	.230	.080	.080	.080
#5	.248	.250	.100	.100	.100
#6	.228	.270	.120	.120	.120
#7	.208	.290	.140	.140	.140
#8	.188	.310	.160	.160	.160
#9	.168	.330	.180	.180	.180
#10	.148	.350	.200	.200	
#11	.220	.220			
#12	.240	.240			
#13	.260	.260			
#14	.280	.280			

Drivers: Chambers 1, 2, 5 and 6 have a stack height of 15.
   Control chambers 3 and 4: stack height is 20. Add 8 to the control key cut and subtract from 20. This is constant throughout the system.

Sargent Interchangeable Core Keying Restrictions

1. Control cut cannot be the same as the corresponding master cut.

2. Control cut and corresponding master cut must be within seven steps of each other.

3. If the master cut is even, corresponding control cut must be even.

4. If the master cut is odd, corresponding control cut must be odd.

5. Change key cuts cannot be the same as control key cuts in both control chambers.

System 70

Shoulder to First Cut: .250"
Center to Center: .156"
Macs: 4

	Root Depth	Bottom Pins	Master Pins	Control Pins	Drivers
#1	.339	.160	.028	−4=.051	0=.247
#2	.311	.189	.056	−3=.080	.192
#3	.283	.217	.084	−2=.107	.163
#4	.255	.245	.112	−1=.135	.135
#5	.227	.273	.140	0=.163	.107
#6	.199	.301		+1=.192	.080
				+2=.218	.080
				+3=.247	
				+4=.275	
				+5=.303	

Non-control chambers use a #0 driver.

In the control chambers the drivers are the same number as the corresponding control key cut. This is constant throughout the system.

System 70 Keying Restrictions

1. No #1 cut allowed in the control position of the control key.

2. The control key must be the same as the master key except for two positions (the two different positions must be in position 2, 3, 4 or 5).

3. The two control cuts that are different may not be used in the system progression.

4. If the system is also using high-security cylinders, the two different positions must be positions 2 and 3. These are the only two control chambers in the high-security cylinders.

Locksmithing Supply Houses/Distributors

Access Hardware Supply
14359 Catalina St.
San Leandro, CA 94577
Phone: 510-483-5000
Toll Free: 1-800-348-2263
Fax: 510-483-4500
Web Site: www.accesshardware.com

Accredited Lock Supply Co.
1161 Peterson Plank Rd.
Secaucus, NJ 07094
Phone: 201-865-5015
Toll Free: 1-800-652-2835
Fax: 201-865-5031
E-Mail: mail@acclock.com
Web Site: www.acclock.com
Officers: Ron Weaver, Pres.

AccuLock Inc.
9901 South IH-35W
Grandview, TX 76050
Phone: 817-866-3918
Toll Free: 1-866-222-8562
Fax: 817-866-3921; 1-800-291-4592
E-Mail: sales@acculock.com
Web Site: www.acculock.com/acculock.biz
Officers: Rick Segerstrom, Pres.

Ace Lock
5964 Baum Blvd.
Pittsburgh, PA 15206
Phone: 412-363-3328
Fax: 412-363-7540
E-Mail: info@acelockinc.com
Web Site: www.acelockinc.com
Officers: Sandra L. Hunter, Pres.;
James Jewell, Sec-Treas.

Ace Security Control Inc.
720 W. Green St.
Ithaca, NY 14850
Phone: 607-273-0526

Acme Lock Inc.
139 E. Court St.
Cincinnati, OH 45202
Phone: 513-241-2614
Fax: 513-241-5612
E-Mail: acmelock@fuse.net
Web Site: www.acmelockinc.com

ADI
263 Old Country Rd.
Melville, NY 11747
Phone: 631-692-1000
Toll Free: 1-800-233-6261;
(nearest location) 1-800-234-7971
E-Mail: info@adi-dist.com
Web Site: www.adilink.com
Officers: Tom Poison, John Sullivan

Adler Video Systems Inc.
711 W. Ivy
Glendale, CA 91204
Phone: 818-409-1701
Toll Free: 1-800-488-7978
Fax: 818-409-0944

Akron Hardware
1100 Killian Rd.
Akron, OH 44312
Toll Free: 1-800-321-9602

Fax: 1-800-328-6070
E-Mail: sales@akronhardware.com
Web Site: www.akronhardware.com
Officers: Kenneth E. Orihel; Tom Orrihel;
Nancy Murray

Akron Hardware (Branch)
3150 N. San Marcos Pl.
Chandler, AZ 85225
Toll Free: 1-800-457-9378
Fax: 1-800-457-8773
E-Mail: azsales@akronhardware.com

Alarm Systems Distributors
883 Broadway
Albany, NY 12207
Phone: 518-463-4000
Toll Free: 1-800-325-6045
Fax: 518-463-4628
Web Site: www.alarmsystemsdist.com
Officers: Sharon Rosen; Ed Rosen

Alarm-Saf
65A Industrial Way
Wilmington, MA 01887-3499
Phone: 978-658-6717
Toll Free: 1-800-987-1050
Fax: 978-658-8638
Web Site: www.alarmsaf.com

Alex Tuttle Co.
3863 NW 19th St.
Ft. Lauderdale, FL 33311
Toll Free: 1-800-417-2002

ALK Contractors Supply Corp.
1266 W. Lake St.
Roselle, IL 60172
Phone: 630-894-0041
Fax: 630-894-8081

Allegheny Distributors
1006–08 Cottman Ave.
Philadelphia, PA 19111
Phone: 215-745-1217
Fax: 215-745-1414

Allied Locksmith Supply
P.O. Box 3137
Youngstown, OH 44513-3137
Phone: 330-726-0866
Toll Free: 1-800-544-2102
Fax: 330-726-0865
E-Mail: wiesnercorp@msn.com
Officers: Rudy R. Wiesner, Pres.; Jim Dravec,
Oper. Mgr.; Bill Stark, Sales Mgr.

ALW Security & Hardware Supply
7020 82nd Ave.
Edmonton, AB T6B OE7 Canada
Phone: 780-465-0184
Toll Free: 1-800-465-5512
Fax: 780-469-8167
Officers: Evan Morgan, Pres.

Ambassador Wholesale Safe Distributors
21 N. Beverwyck Rd.
Lake Hiawatha, NJ 07034
Phone: 973-263-3768
Toll Free: 1-800-999-5510
Fax: 973-402-1447

Amedco Security Systems Inc.
P.O. Box 361295
San Juan, Puerto Rico 00936-1295
Phone: 787-767-2085

American Auto Lock.com
1039 North Christian St.
Lancaster, PA 17602
Phone: 717-392-6333; 717-581-8303
Toll Free: 1-800-860-5625
Fax: 717-581-8353
E-Mail: info@americanautolock.com
Web Site: www.americanautolock.com
Officers: Bill Neff, Pres.;
Barb Neff, VP, Sec/Treas.

American Building Supply
8360 Elder Creek Rd.
Sacramento, CA 98828

Phone: 916-379-4270
Fax: 916-379-4287
E-Mail: sean_deforrest@abs-abs.com
Web Site: www.abs-hardware.com
Officers: Mark Ballantyne, Pres.

Anderson Lock Co. Ltd.
850 E. Oakton St.
Des Plaines, IL 60018
Phone: 847-296-1159
Toll Free: 1-800-323-5625
Fax: 847-296-9259
Web Site: www.andersonlock.com
Officers: Eugene R. Anderson, Courtney
Anderson Washche

Beeson Hardware Co. Inc.
P.O. Box 1390
High Point, NC 27261
Phone: 336-821-2100
Toll Free: 1-800-967-0394
Fax: 336-887-1078
E-Mail: info@beesonhardware.com
Web Site: www.beesonhardware.com

Benton Ltd. J. W.
3491 Main St.
Vancouver, BC V5V 3M9 Canada
Phone: 604-873-0254
Fax: 604-873-6111

Berg Wholesale
P.O. Box 3050
Tualatin, OR 97062
Phone: 503-454-5454
Toll Free: 1-800-243-8887; 1-800-582-8107 (Florida)
Fax: 1-800-343-0937; 503-454-5450
E-Mail: sales@bergwholesale.com
Officers: Catherine Mick

Blaydes Industries Inc.
2335 18th St. N.E.
Washington, D.C. 20018
Phone: 202-832-7100

Toll Free: 1-800-424-2650
Fax: 202-832-1359
E-Mail: blaydesind@msn.com
Web Site: www.blaydesind.com
Officers: E.J. Hildebrand, Pres.; W.T. Rakes, VP;
A. Hildebrand, Sec.

Blue Dog
P.O. Box 160–006
Altamonte Springs, FL 32716
Phone: 407-774-0100
Toll Free: 1-888-ODD-BLANKS
E-Mail: bluedogkeys@earthlink.net
Web Site: www.bluedogkeys.com
Officers: Liam Gribben, Pres.

Boyle & Chase Inc.
72 Sharp St.
Hingham, MA 02043
Phone: 781-337-3904
Toll Free: 1-800-325-2530
Fax: 1-800-205-3500
E-Mail: sales@boyleandchase.com
Web Site: www.boyleandchase.com
Officers: Michael Rotondi, Pres.;
Bob Kent, VP

C.C. Craig Co. Ltd. (Branch)
#10–3419 12th St. NE
Calgary, AB T2E 6S6
Canada
Phone: 403-291-1441
Toll Free: 1-800-565-1242
Fax: 403-291-1573
Officers: Bill Durst, Mgr.

C.C. Craig Co. Ltd. (Branch)
4444 97th St.
Edmonton, AB T6E 5R9 Canada
Phone: 780-409-9504
Toll Free: 1-866-768-0883
Fax: 780-409-9403
Officers: Alan Dickenson, Mgr.

Capitol Lock Co., Inc.
14516 MacClintock Ct.
Glenwood, MD 21738
Phone: 301-937-9500

cctvproducts.com
4444 West Russell Rd., Ste. I
Las Vegas, NV 89118
Phone: 702-367-1525
Toll Free: 1-877-628-2283
E-Mail: info@cctvproducts.com
Web Site: www.cctvproducts.com
Officers: Ron Freschi, Cynthia Freschi

Central Lock & Hardware Supply Co.
95 NW 166th St.
Miami, FL 33169-6048
Phone: 305-947-4853
Toll Free: 1-800-677-4549
Fax: 305-949-8945
E-Mail: sales@cenlock.com
Web Site: www.cenlock.com

Certified Alarm Distributors
4321 18th Ave.
Brooklyn, NY 11218
Phone: 718-435-2800
Fax: 718-435-2818
E-Mail: sales@certifiedalarmdist.com
Web Site: www.certifiedalarmdist.com

Christy Industries Inc.
1812 Bath Ave.
Brooklyn, NY 11214
Phone: 718-236-0211
Toll Free: 1-800-472-2078
Fax: 718-259-3294
E-Mail: tomp@christy-ind.com
Web Site: www.christy-ind.com
Officers: Thomas Principale

City Lock
2898 30th St.
Boulder, CO 80301

Phone: 303-444-4407
Fax: 303-442-8380
E-Mail: lock1234@aol.com

Clark Security Products—Corporate Office

4775 Viewridge Ave.
San Diego, CA 92123
Phone: 858-505-1950
Toll Free: 1-800-854-2088
Fax: 858-495-0081
Web Site: www.clarksecurity.com
Officers: Marshall Merrifield, Pres.; Ginny
Merrifield, VP; Susan Kuruvilla, CFO;
Peter Berg, COO; Peter Cohen, VP Mktg.;
Vicki Wilson, VP National Accounts

Clark Security Products (Anaheim)

1210 N. Kraemer Blvd.
Anaheim, CA 92806
Phone: 714-238-9000
Toll Free: 1-800-889-5625
Fax: 714-238-9088
E-Mail: webmaster@clarksecurity.com
Web Site: www.clarksecurity.com
Officers: Joe Rigby, Reg. Operations Mgr.; Rick Adams, Reg.
Sales Mgr.; Dan Flores, Operations Mgr.; Kirn Omyaian,
Service Mgr.

Clark Security Products (Chicago)

100 Leland Ct., Unit D
Bensenville, IL 60106
Phone: 630-350-8500
Toll Free: 1-800-755-5625
Fax: 630-350-8535
E-Mail: webmaster@clarksecurity.com
Web Site: www.clarksecurity.com
Officers: Kevin Plunkett, Reg. Operations Mgr.; Mary Dover,
Operations Mgr.; Mark Crowley, Service Mgr.; Tanya Alves,
Reg. VP of Sales

Clark Security Products (Dallas)

10390 Shady Trail, Ste. 104
Dallas, TX 75220
Phone: 214-352-3536

Toll Free: 1-800-483-5625
Fax: 214-351-5199
E-Mail: webmaster@clarksecurity.com
Web Site: www.clarksecurity.com
Officers: Kevin Plunkett, Reg. Operations Mgr.;
Randy McKitrick, Reg. Sales Mgr.

Clark Security Products (Denver)
4900 N. Osage #400 Bldg. B
Denver, CO 80221
Phone: 303-288-9200
Toll Free: 1-800-282-5625
Fax: 303-288-7535
E-Mail: webmaster@clarksecurity.com
Web Site: www.clarksecurity.com
Officers: Vicki Wilson, Reg Mgr.; David Wuest, Br. Mgr.;
Wade Sawyer, Service Mgr.

Clark Security Products (Florida)
7576 Kingspointe Pkwy., Ste. 170
Orlando, FL 32819
Phone: 407-226-6770
Toll Free: 1-800-669-5625
Fax: 407-226-6585
E-Mail: webmaster@clarksecurity.com
Web Site: www.clarksecurity.com
Officers: Stephanie Miller, Reg. VP of Operations;
Tony Green, Reg. Operations Mgr.; Tanya Alves, Reg. VP
of Sales; Tom Hightman, Reg. Sales Mgr.

Clark Security Products (Kentucky)
2409 Over Dr.
Lexington, KY 40511
Phone: 859-425-7876
Toll Free: 1-800-659-5625
Fax: 859-425-7872
E-Mail: webmaster@clarksecurity.com
Web Site: www.clarksecurity.com
Officers: Stephanie Miller, Reg. VP of Operations; Tony
Green, Reg. Operations Mgr.; Brent Taylor, Op. Mgr.; Tanya
Alves, Reg. VP of Sales; Tom Hightman, Reg. Sales Mgr.

CMT Inc.
P.O. Box 297
Hamilton, MA 01936
Phone: 978-468-5640
Toll Free: 1-800-659-9140
E-Mail: cmtinc@tiac.net
Web Site: www.habitatmonitor.com

Colonial Lock Supply Co. Inc.
P.O. Box 1417, 7000-G Newington Rd.
Newington, VA 22122
Phone: 703-550-8558
Toll Free: 1-800-732-9117
Fax: 703-550-8857
E-Mail: info@colonialock.com

Commonwealth Lock Co.
1853 Massachusetts Ave.
Cambridge, MA 02140
Phone: 617-876-3301
Toll Free: 1-800-442-7009
Fax: 617-661-3168
E-Mail: commonwealth.lock@verizon.net
Web Site: www.commonwealthlock.com

D and H Distributing Co.
2525 N. 7th St.
P.O. Box 5967
Harrisburg, PA 17110
Phone: 717-255-7866
Toll Free: 1-800-877-1200
Fax: 717-255-7864

D. G. MacLachlan Ltd.
4050 Grant St.
Burnaby, BC V5C 3N5 Canada
Phone: 604-294-6000
Toll Free: 1-800-665-0535
Fax: 604-294-3333
E-Mail: garyr@dgmac.com
Officers: Gary Robertson, Henry Clarke

D & S Products Co., Inc.
9451 Jackson Rd.

Sacramento, CA 95826
Phone: 916-362-3502

D. Silver Hardware Co. Inc.
591 Ferry St.
Newark, NJ 07105
Phone: 201-344-3963
Toll Free: 1-800-222-2915

Dactek International Inc.
8142 Orion Ave.
Van Nuys, CA 91406-1450
Phone: 818-787-1901
Toll Free: 1-800-232-2835
Fax: 818-988-9776
Web Site: www.dactek.com

David Levy Company, Inc.
12753 Moore St.
Cerritos, CA 90703
Phone: 562-404-9998
Toll Free: 1-800-421-3536
Fax: 562-404-9698
E-Mail: alex@dlcparts.com
Web Site: www.dlcparts.com
Officers: David Levy, John Latino,
Gordon Shear

Deutschen & Sons
105-07-150th St.
Jamaica, NY 11435
Phone: 718-291-5600

Diebold Inc.
5995 Mayfair Rd.
North Canton, OH 44720
Phone: 330-490-4000
Toll Free: 1-800-999-3600
E-Mail: productinfo@diebold.com
Web Site: www.diebold.com

Dire's Lock Co. Inc.
2201 Broadway
Denver, CO 80205
Phone: 303-294-0176

Fax: 303-294-0198
Officers: Donna Dire, Pres.; Michael Dire, VP

Edward Saucedo & Son Co. Inc.
711 N. Copia St.
El Paso, TX 79903-4405
Phone: 915-566-7101
Fax: 915-566-8608
E-Mail: esscoinc@hotmail.com
Officers: De Saucedo, GM

Elbex America Inc.
10761 Noel St.
Los Alamitos, CA 90720
Phone: 714-761-8000
Toll Free: 1-800-367-2288
Fax: 714-761-8400
E-Mail: elbex@att.net
Web Site: www.elbex-video.com

Eljay Express
P.O. Box 1388
Wheeling, IL 60090
Phone: 847-541-3343
Fax: 847-541-3363

EMG Associates
645 N. Michigan Ave.
Chicago, IL 60611
Phone: 312-649-0662
Toll Free: 1-800-468-3558
Fax: 312-649-0787

Empire Safe Co. Inc.
6 E. 39th St.
New York, NY 10016
Phone: 212-684-2255
Toll Free: 1-800-543-5412
Fax: 212-684-5550
E-Mail: empiresafe@aol.com
Web Site: www.empiresafe.com
Officers: Richard Krasilovsky

Empire Security Supplies
4600 B Nesconset Hwy.

Port Jefferson Station, NY 11776
Phone: 631-928-1919
Fax: 631-928-4745

Engineered Security Systems Inc.
1 Indian Lane East
Towaco, NJ 07034
Phone: 973-257-0555
Toll Free: 1-800-742-1263
Fax: 973-257-0550
Web Site: engineeredsecurity.com

Ewert Wholesale Hardware Inc.
5801 W. 117th Place
Alsip, IL 60658
Phone: 708-597-0059
Fax: 708-597-0881
Officers: Clelia A. King, Pres.;
A. Wesley McKenney, GM

Fairway Supply Inc.
8222 N. Lamar, Ste. F53
Austin, TX 78753
Phone: 512-452-6300
Toll Free: 1-877-309-6300
Fax: 512-452-8014

Fastec Industrial Corp.
23348 CT Rd. 6
Elkhart, IN 46514
Phone: 574-262-2505
Toll Free: 1-800-837-2505
Fax: 574-266-0123
E-Mail: chuckrw@fastecindustrialcorp.com
Web Site: fastecindustrialcorp.com
Officers: Charles R. White

Foley-Belsaw Co./Locksmith Supply Div.
1760 Universal Ave.
Kansas City, MO 64120
Phone: 816-483-6400
Toll Free: 1-800-821-3452
Fax: 816-483-5010
Web Site: www.foley-belsaw.com

Fowler of Canada
24 E. McIntyre Pl.
Kitchener, ON N2R 1G9 Canada
Phone: 519-895-0970

Franklin Mfg., Division of Florida Pneumatic Mfg. Group
851 Jupiter Park Lane
Jupiter, FL 33458
Toll Free: 1-800-488-5304
Fax: 1-800-488-5316
E-Mail: Lfranklin@franklinmfg.com
Web Site: www.franklinmfg.com
Officers: Larry Franklin, Jesse Taub

Fried Bros. Inc.
467 N. 7th St.
Philadelphia, PA 19123
Phone: 215-627-3205
Toll Free: 1-800-523-2924
Web Site: www.FBIsecurity.com
Officers: Alex Ebrahimzadeh, Pres.

Gil-Ray Tools Inc.
P.O. Box 801, 1306 McGraw St.
Bay City, MI 48708
Phone: 989-892-6870
Fax: 989-892-6870
E-Mail: djdeuel@bigfoot.com
Web Site: www.angelfire.com/biz/
GilRayToolsInc/
Officers: David Deuel, President

Goodbar Security
1100 11th St. SE
Calgary, AB T2G 4T3 Canada
Phone: 403-234-9690
Toll Free: 1-800-661-7555
Fax: 403-237-5894
Web Site: www.goodbarsecurity.com
Officers: Tricia Radison, VP Mktg.

Granite Security Products Inc.
4801 Esco Dr.
Fort Worth, TX 76140

Phone: 817-561-9095
Fax: 817-478-3056
Web Site: www.granitesafe.com

Hardinge Bros. Ltd.
1-3375 14th Ave.
Markham, ON L3R OH2 Canada
Phone: 905-470-8844
Toll Free: 1-800-387-5065
Fax: 905-470-2767
E-Mail: info@hardingecansecure.com
Web Site: www.hardingecansecure.com
Officers: Kevin Greer, Natasha Richardson

Hardware Agencies Ltd.
1220 Dundas St. E.
Toronto, ON M4M 132 Canada
Phone: 416-462-1919
Toll Free: 1-800-268-6741
Fax: 1-800-903-3303
E-Mail: sales@hardwareagencies.com
Web Site: www.hardwareagencies.com

Hawley Lock Supply
2915 E. Poinsettia Dr.
Phoenix, AZ 85028
Phone: 602-795-8247
Toll Free: 1-800-398-2458
Fax: 602-795-8047
E-Mail: hawleylock@gmail.com
Web Site: www.hawleylock.com

High Tech Tools
1628 NW 28th St.
Miami, FL 33142
Phone: 305-635-1011
Toll Free: 1-800-323-8324
Fax: 305-635-1015
E-Mail: sales@hightechtools.com
Web Site: www.hightechtools.com
Officers: Tony Vigil

Howard Sales Co.
4625 Ripley Dr.

El Paso, TX 79922
Phone: 915-833-7733
Toll Free: 1-800-456-4625
Fax: 915-833-7770
E-Mail: info@howardsales.com
Web Site: www.howardsales.com
Officers: Bill Uhlig, Pres.

Hy-Security Gate Operators
1200 W. Nickerson
Seattle, WA 98119
Phone: 206-286-0933
Toll Free: 1-800-321-9947
Fax: 206-286-0614
E-Mail: sales@hy-security.com
Web Site: www.hy-security.com
Officers: Brian Denault, VP

IDN—International Distribution Network Inc.
2401 Mustang Dr. Ste. 100
Grapevine, TX 76051
Phone: 817-421-5470
Fax: 817-421-5468
Web Site: www.idn-inc.com
Officers: Mike Groover, Pres.; Al Hoffman, VP;
Karen Hoffman Kahl, Sec./Treas.

IDN—International Distribution Network Inc.
(Branches)
Atlanta, GA Phone: 404-875-0136
Toll Free: 1-800-726-3332

Albuquerque, NM Phone: 505-816-0100
Toll Free: 1-800-291-1222

Buffalo, NY Phone: 716-626-1208
Toll Free: 1-888-588-5768

Calgary, AB Canada Phone: 403-291-4844
Toll Free: 1-800-813-5344

Chicago, IL Phone: 708-456-9600
Toll Free: 1-800-433-6608

Cincinnati, OH Phone: 513-271-8530
Toll Free: 1-800-457-3963

Cleveland, OH Phone: 216-335-9740
Toll Free: 1-800-247-8217

Columbus, OH Phone: 614-436-6619
Toll Free: 1-800-774-3255

Dallas, TX Phone: 972-664-1240
Toll Free: 1-800-372-2263

Davenport, IA Phone: 563-391-8366
Toll Free: 1-800-3914633

Denver, CO Phone: 303-922-3041
Toll Free: 1-800-445-4008

Detroit, MI Phone: 586-755-3658
Toll Free: 1-800-468-7490

Edmonton, AB Canada Phone: 780-944-0014

Ft. Worth, TX Phone: 817-284-5696
Toll Free: 1-800-859-2263

Grand Rapids, MI Phone: 616-534-1067
Toll Free: 1-800-244-0777

Halifax, NS Canada Phone: 902-468-1373

Harrisburg, PA Phone: 717-909-9740

Hartford, CT Phone: 860-296-7886
Toll Free: 1-877-477-5625

Indianapolis, IN Phone: 317-635-8100
Toll Free: 1-800-428-9313

Jacksonville, FL Phone: 904-387-0663
Toll Free: 1-800-341-7857

Kansas City, KS Phone: 913-599-4111
Toll Free: 1-800-526-5897

Las Vegas, NV Phone: 702-736-4553
Toll Free: 1-800-556-9388

Livonia, MI (three locations) Phone: 734-591-1150
Toll Free: 1-800-521-0955

Los Angeles, CA Phone: 562-463-4870
Toll Free: 1-800-203-1577

Memphis, TN Phone: 901-795-2250
Toll Free: 1-800-687-1263

Miami, FL Phone: 305-651-1598
Toll Free: 1-800-827-3332

Milwaukee, WI Phone: 262-790-9750
Toll Free: 800-882-9899

Minneapolis, MN Phone: 952-886-3840
Toll Free: 1-800-326-8986

Montreal, PQ Canada Phone: 514-956-0248

New Orleans, LA Phone: 504-837-7315
Toll Free: 1-800-788-2263

Norfolk, VA Phone: 757-853-0611
Toll Free: 1-800-735-3334

Oklahoma City, OK Phone: 405-942-8750
Toll Free: 1-800-664-2263

Omaha, NE Phone: 402-592-1652
Toll Free: 1-800-824-0684

Orlando, FL Phone: 407-297-7722
Toll Free: 1-800-775-1220

Ottawa, ON Canada Phone: 613-749-2172
Toll Free: 1-800-550-4209

Philadelphia, PA Phone: 215-289-4266
Toll Free: 1-800-233-3355

Phoenix, AZ Phone: 602-272-5300
Toll Free: 1-800-525-3131

Pittsburgh, PA Phone: 412-771-6122
Toll Free: 1-800-837-5625

Raleigh, NC Phone: 919-277-0007
Toll Free: 1-800-673-3330

Riverside, CA Phone: 951-788-8300
Toll Free: 1-800-203-1577

San Antonio, TX Phone: 210-545-3396
Toll Free: 1-800-431-2263

Syracuse, NY Phone: 315-455-9305
Toll Free: 1-800-805-0408

St. Louis, MO Phone: 314-692-8391
Toll Free: 1-800-392-1303

Tampa, FL Phone: 813-886-8007
Toll Free: 1-800-829-3332

Toronto, ON Canada Phone: 416-248-5625
Toll Free: 1-800-268-1306

Tucson, AZ Phone: 520-322-5625

Vancouver BC Canada Phone: 604-253-0017
Toll Free: 1-800-567-1177

IDN—Acme, Inc.
P.O. Drawer 13748
New Orleans, LA 70185
Phone: 504-837-7315
Toll Free: 1-800-788-2263
Fax: 504-837-7321
Web Site: www.idnacme.com
Officers: Barry Johnson, Pres.

IDN—Acme, Inc. (Branches)
Albuquerque, NM Phone: 505-816-0100
Toll Free: 1-800-291-1222 Fax: 505-816-0111

Dallas, TX Phone: 972-664-1240
Toll Free: 1-800-372-2263 Fax: 972-664-1252

Denver, CO Phone: 303-992-3041
Toll Free: 1-800-445-4008 Fax: 303-922-7493

Ft. Worth, TX Phone: 817-284-5696
Toll Free: 1-800-859-2263 Fax: 817-284-3501

IDN—Armstrong's, Inc.
3589 Broad St.
Chamblee, GA 30341
Phone: 404-875-0136
Toll Free: 1-800-726-3332
Fax: 404-888-0834
Web Site: www.idnarmstrongs.com
Officers: John Burke, Pres.

IDN—Armstrong's, Inc. (Branches)
Jacksonville, FL Phone: 904-387-0663
Toll Free: 1-800-341-7857 Fax: 904-387-4568

Miami, FL Phone: 305-652-2598
Toll Free: 1-800-827-3332 Fax: 305-652-1709

Norfolk, VA Phone: 757-853-0611
Toll Free: 1-800-735-3334 Fax: 757-853-1832

Orlando, FL Phone: 407-297-7722
Toll Free: 1-800-775-1220 Fax: 407-297-9522

Raleigh, NC Phone: 919-277-0007
Toll Free: 1-800-673-3330 Fax: 919-277-0014

Tampa, FL Phone: 813-886-8007
Toll Free: 1-800-829-3332 Fax: 813-886-5696

IDN—Hardware Sales, Inc.
35950 Industrial Rd.
Livonia, MI 48150
Phone: 734-591-1150
Toll Free: 1-800-521-0955
Fax: 734-591-3981
Web Site: www.idnhardware.com
Officers: Arnie Goldman, Pres.

IDN—Hardware Sales, Inc. (Branches)
Buffalo, NY Phone: 716-626-1208
Toll Free: 1-888-588-5768

Cleveland, OH Phone: 216-335-9740
Toll Free: 1-800-247-8217

Detroit, MI Phone: 586-755-3658
Toll Free: 1-800-468-7490

Harrisburg, PA Phone: 717-909-9740

Hartford, CT Phone: 860-296-7886
Toll Free: 1-877-477-5625

Philadelphia, PA Phone: 215-289-4266
Toll Free: 1-800-233-3355

Pittsburgh, PA Phone: 412-771-6122
Toll Free: 1-800-837-5625

Syracuse, NY Phone: 315-455-9305
Toll Free: 1-800-805-0408

IDN—H. Hoffman, Inc.
7330 West Montrose Ave.
Chicago, IL 60706
Phone: 708-456-9600
Toll Free: 1-800-433-6608
Fax: 708-456-0878
Web Site: www.idnhoffman.com
Officers: Karen Hoffman Kahl, Pres.

IDN—H. Hoffman, Inc. (Branches)
Cincinnati, OH Phone: 513-271-8530
Toll Free: 1-800-457-3963

Columbus, OH Phone: 614-436-6619
Toll Free: 1-800-774-3255
Davenport, IA Phone: 563-391-8366
Toll Free: 1-800-3914633

Indianapolis, IN Phone: 317-635-8100
Toll Free: 1-800-428-9313

Kansas City, KS Phone: 913-599-4111
Toll Free: 1-800-526-5897

Milwaukee, WI Phone: 262-790-9750
Toll Free: 800-882-9899

Minneapolis, MN Phone: 952-886-3840
Toll Free: 1-800-326-8986

Omaha, NE Phone: 402-592-1652
Toll Free: 1-800-824-0684

St. Louis, MO Phone: 314-692-8391
Toll Free: 1-800-392-1303

IDN—Canada, Ltd.
70 Floral Parkway
Toronto, ON M6L 2B9 Canada
Phone: 416-248-5625
Toll Free: 1-800-268-1306
Fax: 416-248-9945
Web Site: www.idn-canada.com
Officers: Steve Dyson, Pres.

IDN—Canada, Ltd. (Branches)
Calgary, AB Canada Phone: 403-291-4844
Toll Free: 1-800-813-5344

Edmonton, AB Canada Phone: 780-944-0014

Halifax, NS Canada Phone: 902-468-1373

Montreal, PQ Canada Phone: 514-956-0248

Ottawa, ON Canada Phone: 613-749-2172
Toll Free: 1-800-550-4209

Vancouver BC Canada Phone: 604-253-0017
Toll Free: 1-800-567-1177

IDN—West, Inc. (Branches)
Las Vegas, NV Phone: 702-736-4553
Toll Free: 1-800-556-9388

Los Angeles, CA Phone: 562-463-4870
Toll Free: 1-800-203-1577

Riverside, CA Phone: 951-788-8300
Toll Free: 1-800-203-1577

Tucson, AZ Phone: 520-322-5625

Independent Hardware Inc.
14 S. Front St.
Philadelphia, PA 19106-3001
Phone: 215-925-5306
Toll Free: 1-800-346-9464
Fax: 215-922-2887; 215-922-6552
E-Mail: sales@independenthardware.com
Web Site: www.independenthardware.com
Officers: Frank Stanco, Pres.; Vincent Campagna, VP

Inout Systems Inc.
3650-B Matte Blvd.
Brossard, PQ J4Y 2Z2 Canada
Phone: 450-444-5949
Fax: 450-444-4856
Web Site: www.inoutsystems.com

Intermountain Lock & Security Supply (Las Vegas)
4545 Cameron St., Suite A
Las Vegas, NV 89103
Phone: 702-939-5625
Toll Free: 1-866-809-5625
Fax: 702-939-5626
E-Mail: nealh@intermountainlock.com
Web Site: www.imlss.com

Intermountain Lock & Security Supply (Los Angeles)
5950 Raster St.
Los Angeles, CA 91411
Phone: 818-781-9799
Toll Free: 1-800-729-5444
Fax: 818-781-1828
E-Mail: ellenj@intermountainlock.com
Web Site: www.imlss.com

Intermountain Lock & Security Supply (Salt Lake City)
3106 S. Main St.
Salt Lake City, UT 84115

Phone: 801-486-0079
Toll Free: 1-800-453-5386
Fax: 801-485-7205
Web Site: www.imlss.com

International Door Closers
1920 Air Lane Dr.
Nashville, TN 37210
Phone: 615-885-7060
Toll Free: 1-800-225-6737
Fax: 615-885-0903
E-Mail: idcten@intldoorclosers.com
Web Site: www.intldoorclosers.com

Island Pacific Distributors Inc.
240-A Puuhale Rd.
Honolulu, HI 96819-2262
Phone: 808-955-1126
Fax: 808-946-6480
E-Mail: ipd@aloha.net

J & M Distributing
P.O. Box 18-032
Chicago, IL 60618-0032
Phone: 773-583-8787
Toll Free: 1-800-645-0040
Fax: 773-583-0040
E-Mail: rjakubowski@ameritech.net
Officers: Richard Jakubowski, Jack Metcalf

J. Ross Boles Co. Inc.
606 Broadway
San Antonio, TX 78215
Phone: 210-226-5351
Fax: 210-226-0024
E-Mail: info@jrossbolessafes.com
Web Site: www.jrossbolessafes.com

Jensen Tools Inc./The Stanley Works
7815 S. 46th St.
Phoenix, AZ 85044
Phone: 602-453-3169
Toll Free: 1-800-426-1194
Fax: 602-438-1690; 800-366-9662

E-Mail: jensen@stanleyworks.com
Web Site: www.jensentools.com
Officers: Kai Juel, Pres.

JLM Wholesale
3095 Mullins Ct.
Oxford, MI 48371
Phone: 248-628-6440
Toll Free: 1-800-522-2940; 1-877-212-6203
Fax: 877-212-6202; 1-800-782-1160

JoVan Distributors Inc.
2350 Midland Ave.
Toronto, ON MIS 1P8 Canada
Phone: 416-288-6306
Toll Free: 1-888-752-7210
Fax: 416-752-8371
E-Mail: lmalo@jovanlock.com
Web Site: www.jovanlock.com
Officers: Steve Zizzoz

Johnson/CAAL Enterprises Inc.
8120 Pinewood Circle
Chaska, MN 55317
Phone: 612-448-9550
Fax: 612-448-7898

JS Products Inc.
5440 S. Procyon Ave.
Las Vegas, NV 89135
Phone: 702-212-1343; 702-362-7011
Fax: 702-362-5084
E-Mail: dbuckner@steelman-js.com
Web Site: www.steelman-js.com

Kenstan Lock Company
101 Commercial St.
Plainview, NY 11803
Phone: 516-576-9090
Toll Free: 1-800-859-RUSH
Fax: 516-576-0100
E-Mail: rharrison@kenstan.com
Web Site: www.kenstan.com

Key Products
2555 International St.
Columbus, OH 43228
Phone: 614-529-5801
Toll Free: 1-800-457-1019
Fax: 1-800-435-9054
E-Mail: markw@keyproducts.com
Web Site: www.keyproducts.com
Officers: Matthew J. Finn

Key Sales & Supply Co. Inc.
9950 Freeland Ave.
Detroit, MI 48227
Phone: 313-931-7720
Toll Free: 1-800-445-5397
Fax: 313-931-7758; 800-828-5397
Officers: Michael Wiener, Pres.

Key Supply
362 7th St.
San Francisco, CA 94103
Phone: 415-626-2526
Fax: 415-626-4598

Keys Plus Distributing
454 Young St.
Tonawanda, NY 14150
Phone: 716-743-0220
Toll Free: 1-800-644-5397

Keys Wholesale Distributors Inc.
P.O. Box 498
Drexel Hill, PA 19026-0498
Phone: 610-626-4787
Toll Free: 1-800-292-5397
Fax: 610-626-4189
E-Mail: keys-usa@att.net
Officers: William Kline, Pres.; Dorothy Kline, Sec./Treas.

Kramer Boys Locksmith Supply
1312 S. Broad St.
Trenton, NJ 08610
Phone: 609-393-0707
Toll Free: 1-800-222-2692

Fax: 609-586-0512
E-Mail: kramrboys@aol.com
Web Site: www.kramerboys.com
Officers: Robert Kramer, Scott Kramer, Mike Pororsky

LDM Enterprises
30420 Canwood St., Suite 134
Agoura Hills, CA 91301
Toll Free: 1-800-451-5950
Fax: 818-706-8560
E-Mail: ldment.com@pacbell.net
Web Site: www.ldment.com

Liebert Corp.
1050 Dearborn Dr.
P.O. Box 29186
Columbus, OH 43229
Phone: 614-888-0246
Toll Free: 1-800-877-9222
Fax: 614-841-6022
Web Site: www.liebert.com

Lock Technology Inc.
552 S. Washington St., Suite 108
Naperville, IL 60540
Phone: 630-369-6060
Toll Free: 1-800-421-7421
Fax: 630-983-5967

LockPicks.com
P.O. Box 1900
San Jose, CA 95109
Phone: 408-437-0505
Toll Free: 1-800-480-0875
Fax: 408-437-3129
E-Mail: info@lockpicks.com
Web Site: www.lockpicks.com
Officers: Barrett Brockage, Pres. & CEO

Locks Co.
2050 NE 151st St.
N. Miami, FL 33162
Phone: 305-949-0700
Toll Free: 1-800-288-0801

Fax: 305-949-3619
E-Mail: miami@locksco.com
Web Site: www.locksco.com
Officers: Mark Dorn; Milton Dorn, Pres.; Michael Dorn

Locksmith School Inc.

3901 S. Meridian St.
Indianapolis, IN 46217
Phone: 317-632-3979
Fax: 317-784-2945
E-Mail: jimw@locksmithschoolinc.com
Web Site: www.locksmithschoolinc.com

Locksmith Store Inc., The

1229 E. Algonquin Rd., Unit E
Arlington Heights, IL 60005
Phone: 847-364-5111
Fax: 847-364-5125
E-Mail: info@locksmithstore.com
Web Site: www.locksmithstore.com

LV Sales Inc.

1831 Hyperion Ave.
Los Angeles, CA 90027-4745
Phone: 323-661-4746
Toll Free: 1-800-894-KEYS
Fax: 323-661-1314
E-Mail: sales@lvsales.com
Web Site: www.lvsales.com
Officers: Bernard C. Poulin, CEO

M-D Building Material Co.

953 Seton Ct.
Wheeling, IL 60090
Phone: 847-541-0002
Toll Free: 1-800-345-7116
Fax: 847-541-0003
E-Mail: info@mdhardware.com
Web Site: www.mdhardware.com
Officers: Ralph Menn, Pres.; David Menn, VP; Andy
Dellamaria, Cat. Mgr.; Larry Luxem, Contract Mgr.

Majestic Lock Co. Inc.
65 Liliarts Lane
Elmwood Park, NJ 07407
Phone: 201-475-8400
Fax: 201-475-8484
E-Mail: mike@majesticlockusa.com
Web Site: www.majesticlockusa.com

Malcom Company
25 Abbot St.
Andover, MA 01810
Phone: 978-474-0335
Toll Free: 1-800-289-7505
Fax: 978-475-7242

Mancini Safe Co. Inc.
180 D Kerry Pl.
Norwood, MA 02062
Phone: 781-255-0411
Toll Free: 1-800-367-3453
Fax: 781-255-0082
E-Mail: mancinisafes@erols.com
Web Site: www.mancinisafes.com
Officers: Robert E. Mancini, Pres.

Martco
337 Byrne Ave.
Louisville, KY 40209
Phone: 502-635-1600
Toll Free: 1-800-233-7769
Fax: 502-637-6430
Web Site: www.martcoinc.com
Officers: Gary Armstrong, Dir. Mktg.

MasterLink Security Products Inc.
3863 NW 19th St.
Fort Lauderclale, FL 33311
Toll Free: 1-877-347-9669
E-Mail: masterlink@mindspring.com

Mayflower Sales Co. Inc.
614 Bergen St.
Brooklyn, NY 11238
Phone: 718-622-8785

Toll Free: 1-800-221-2052
Fax: 718-789-8346
Web Site: www.mfsales.com
Officers: Paul Swetow, Bill Swetow

Maziuk Wholesale Distributors
1251 W. Genesee St.
Syracuse, NY 13204
Phone: 315-474-3959, 315-474-3955
Toll Free: 1-800-777-5945, 1-888-605-8311
Fax: 315-472-3111
E-Mail: sales@maziuk.com
Web Site: www.maziuk.com
Officers: Stanley J. Maziuk, Pres.

McDonald Dash Locksmith Supply Inc.
5767 E. Shelby Dr.
Memphis, TN 38141
Phone: 901-797-8000
Toll Free: 1-800-238-7541
Fax: 901-366-0005
E-Mail: jim@mcdonalddash.com
Web Site: www.mcdonalddash.com
Officers: Jim Thomas, Pres.

Vision Quest
251 S. Union
Springfield, MO 65802
Phone: 417-862-1967
Toll Free: 1-800-284-4140
Fax: 417-869-8835
E-Mail: info@vqcctv.com
Web Site: www.vqcctv.com
Officers: Larry Welbern, Lea Hilton

Wespac Corp.
342 Harriet St.
San Francisco, CA 94103-4716
Phone: 415-431-6350
Fax: 415-864-1417; 415-621-1525
E-Mail: wespaccorp@earthlink.net
Web Site: www.wespaccorp.com

Wholesale 4 Inc.
706 S.E. Grand Ave.

Portland, OR 97214
Phone: 503-238-8605
Toll Free: 1-800-547-0921
Fax: 503-813-9290
E-Mail: contact@wholesale4inc.com
Web Site: www.wholesale4inc.com
Officers: Lisa Crawford, Pres.

Wilco Supply
5960 Telegraph Ave.
Oakland, CA 94609
Phone: 510-652-8522
Toll Free: 1-800-745-5450
Fax: 510-653-5397; 1-800-876-5397
E-Mail: info@wilcosupply.com
Web Site: www.wilcosupply.com
Officers: H.D. Williams, Pres.; Robert. D. Williams, VP;
Jon Clarner, Sec./Treas.

Williams Key Co. Inc.
2206 Locust St.
St. Louis, MO 63103
Phone: 314-231-2411
Toll Free: 1-800-325-1779
Fax: 314-231-4609

Wilson Safe Company
3031 Island Ave., Box 5310
Philadelphia, PA 19142
Phone: 215-492-7100; 516-739-2557
Toll Free: 1-800-345-8053
Fax: 215-492-7104; 516-739-5455
E-Mail: RKassoff@WilsonSafe.com
Web Site: www.WilsonSafe.com
Officers: Ray Wilson

Wm. B. Allen Supply Co. Inc.
301 N. Rampart St.
New Orleans, LA 70112
Phone: 504-525-8222
Toll Free: 1-800-535-9593
Fax: 504-525-6361
E-Mail: info@wmballen.com
Web Site: www.wmballen.com

Yates & Felts Inc.
17890 S. Ideal Pkwy.
Manteca, CA 95336
Toll Free: 1-888-692-0404; 1-888-692-2573
E-Mail: bfinney@yatesandfelts.com
Web Site: www.yatesandfelts.com

Zipf Lock Co.
830 Harmon Ave.
Columbus, OH 43223
Phone: 614-228-3507
Toll Free: 1-800-848-1577
Fax: 614-228-6320; 1-800-228-6320
E-Mail: zipflockco@aol.com
Web Site: www.zipflockco.com

D

Electronic Schematics

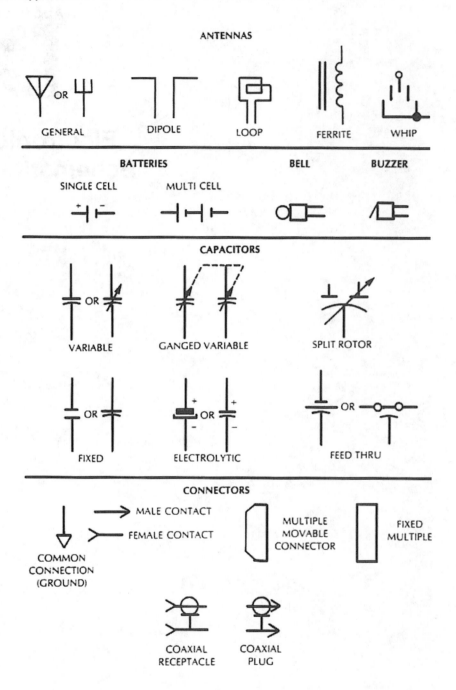

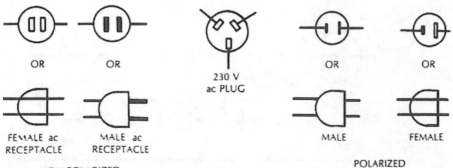

OR   OR   OR   OR

230 V
ac PLUG

FEMALE ac   MALE ac   MALE   FEMALE
RECEPTACLE   RECEPTACLE

NON-POLARIZED   POLARIZED

CRYSTALS

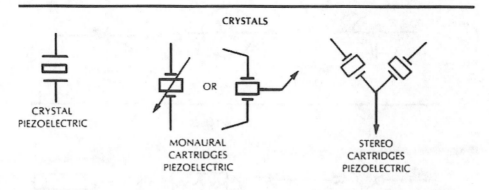

CRYSTAL
PIEZOELECTRIC

OR

MONAURAL
CARTRIDGES
PIEZOELECTRIC

STEREO
CARTRIDGES
PIEZOELECTRIC

DIODES

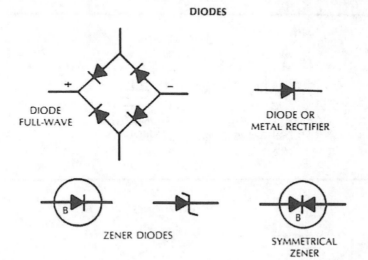

+   −

DIODE
FULL-WAVE

DIODE OR
METAL RECTIFIER

B

ZENER DIODES

B

SYMMETRICAL
ZENER

DIODES (cont'd)

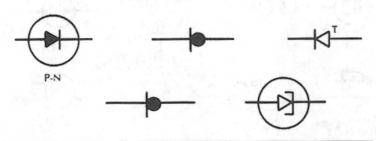

P-N

TUNNEL DIODES

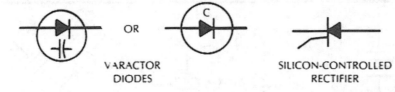

VARACTOR
DIODES

SILICON-CONTROLLED
RECTIFIER

FUSES

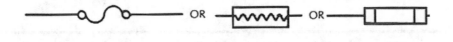

GROUNDS

CHASSIS GROUND
(NOT NECESSARILY
AT GROUND)

EARTH
GROUND

HEADPHONES

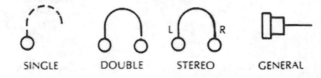

SINGLE   DOUBLE   STEREO   GENERAL

READOUT INDICATOR

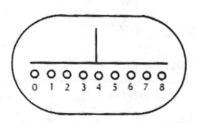

INDUCTORS (COILS)

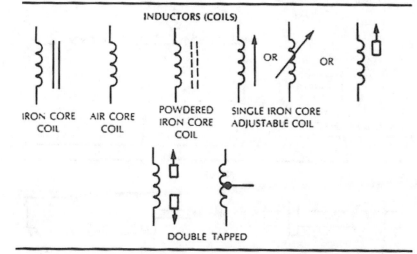

IRON CORE
COIL

AIR CORE
COIL

POWDERED
IRON CORE
COIL

SINGLE IRON CORE
ADJUSTABLE COIL

OR   OR

DOUBLE TAPPED

JACKS

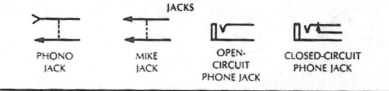

PHONO
JACK

MIKE
JACK

OPEN-
CIRCUIT
PHONE JACK

CLOSED-CIRCUIT
PHONE JACK

LAMPS

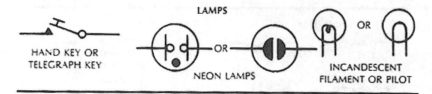

HAND KEY OR
TELEGRAPH KEY

OR

NEON LAMPS

OR

INCANDESCENT
FILAMENT OR PILOT

METERS

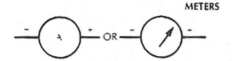

OR

A = AMMETER
mA = MILLIAMMETER
V = VOLTMETER
dB = DECIBEL METER

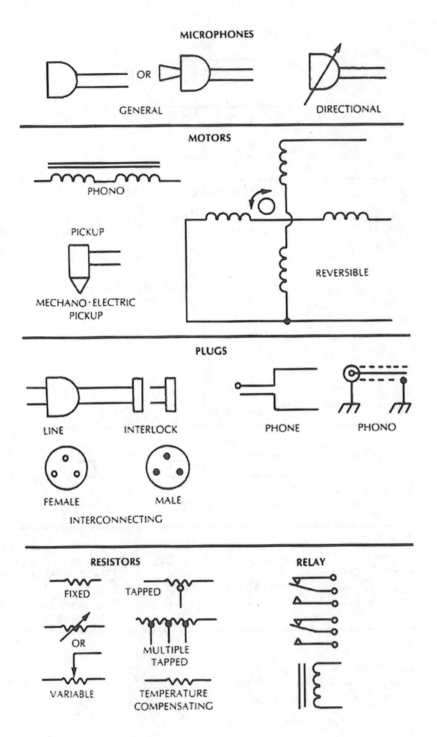

MICROPHONES

OR

GENERAL

DIRECTIONAL

MOTORS

PHONO

PICKUP

MECHANO-ELECTRIC
PICKUP

REVERSIBLE

PLUGS

LINE   INTERLOCK

PHONE   PHONO

FEMALE   MALE

INTERCONNECTING

RESISTORS

FIXED   TAPPED

OR

VARIABLE

MULTIPLE
TAPPED

TEMPERATURE
COMPENSATING

RELAY

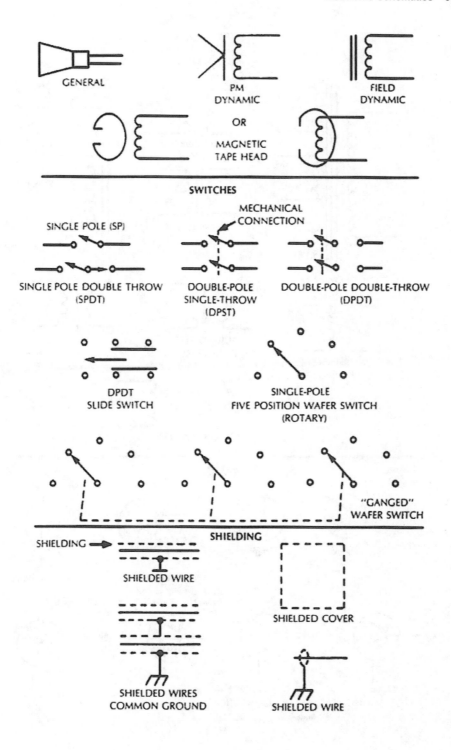

GENERAL

PM
DYNAMIC

FIELD
DYNAMIC

OR

MAGNETIC
TAPE HEAD

SWITCHES

SINGLE POLE (SP)

MECHANICAL
CONNECTION

SINGLE POLE DOUBLE THROW
(SPDT)

DOUBLE-POLE
SINGLE-THROW
(DPST)

DOUBLE-POLE DOUBLE-THROW
(DPDT)

DPDT
SLIDE SWITCH

SINGLE-POLE
FIVE POSITION WAFER SWITCH
(ROTARY)

"GANGED"
WAFER SWITCH

SHIELDING

SHIELDING

SHIELDED WIRE

SHIELDED COVER

SHIELDED WIRES
COMMON GROUND

SHIELDED WIRE

TRANSFORMERS

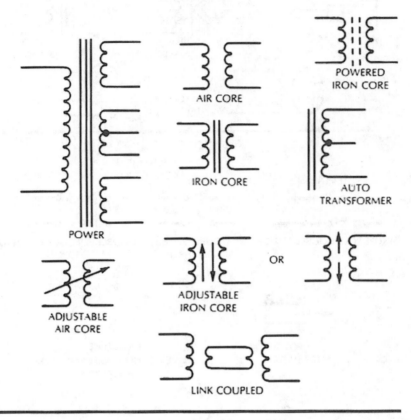

POWERED
IRON CORE

AIR CORE

IRON CORE

AUTO
TRANSFORMER

POWER

ADJUSTABLE
AIR CORE

ADJUSTABLE
IRON CORE

OR

LINK COUPLED

TRANSISTORS

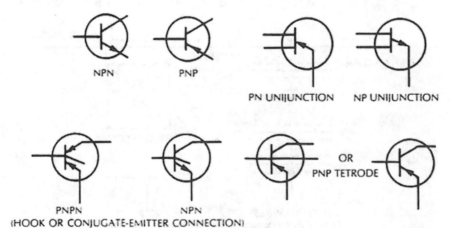

NPN   PNP

PN UNIJUNCTION   NP UNIJUNCTION

PNPN
(HOOK OR CONJUGATE-EMITTER CONNECTION)

NPN

OR
PNP TETRODE

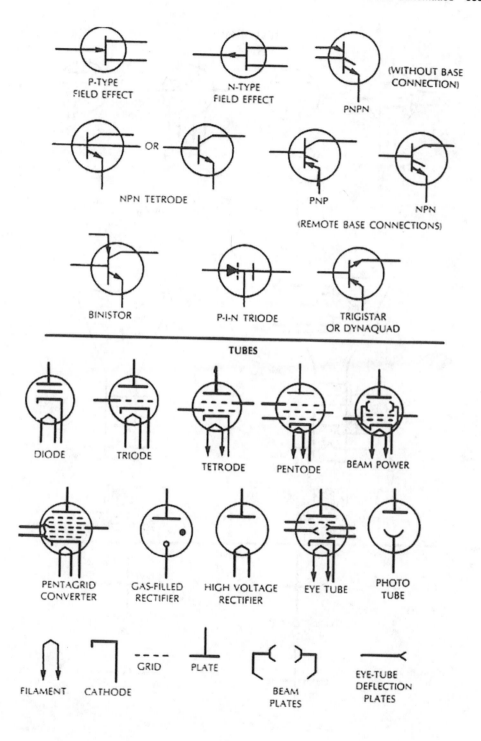

P-TYPE FIELD EFFECT

N-TYPE FIELD EFFECT

PNPN (WITHOUT BASE CONNECTION)

NPN TETRODE — OR

PNP

NPN

(REMOTE BASE CONNECTIONS)

BINISTOR

P-I-N TRIODE

TRIGISTAR OR DYNAQUAD

TUBES

DIODE

TRIODE

TETRODE

PENTODE

BEAM POWER

PENTAGRID CONVERTER

GAS-FILLED RECTIFIER

HIGH VOLTAGE RECTIFIER

EYE TUBE

PHOTO TUBE

FILAMENT

CATHODE

GRID

PLATE

BEAM PLATES

EYE-TUBE DEFLECTION PLATES

TUBES (cont'd)

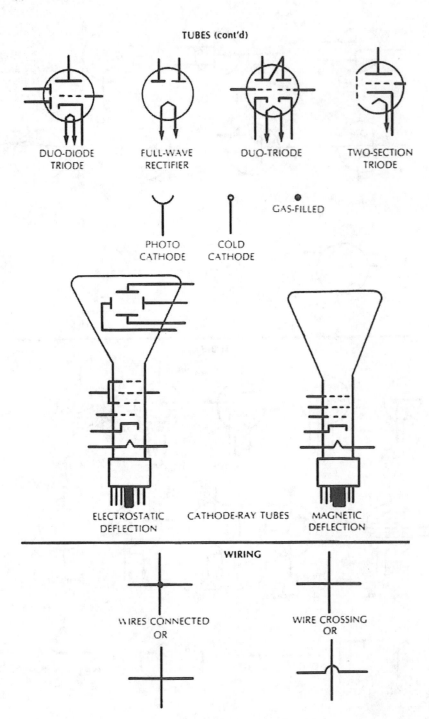

DUO-DIODE
TRIODE

FULL-WAVE
RECTIFIER

DUO-TRIODE

TWO-SECTION
TRIODE

PHOTO
CATHODE

COLD
CATHODE

GAS-FILLED

ELECTROSTATIC
DEFLECTION

CATHODE-RAY TUBES

MAGNETIC
DEFLECTION

WIRING

WIRES CONNECTED
OR

WIRE CROSSING
OR

Central
Monitoring Stations

A-1 Security, Ltd.
917 S. First St.
Las Vegas, NV 89101
Phone: 702-474-6667
Fax: 702-474-9340

AAA Alarm Systems, Ltd.
180 Nature Park Way
Winnipeg, Manitoba R3P 0X7
Canada
Phone: 204-949-0078
Fax: 204-949-9139

Alarm Center, Inc.
P.O. Box 3401
Lacey, WA 98509-3401
Phone: 360-456-1441
Fax: 360-438-4245

Alarm Central LLC
5510 E. 31st St.
Kansas City, MO 64128-1820
Phone: 888-463-6023
Fax: 816-861-4247

Alarm Monitoring Services
200 Desiard St.
Monroe, LA 71201-7334
Phone: 318-398-3301; 877-749-0283
Fax: 504-456-8737

American Alarm & Communications, Inc.
297 Broadway
Arlington, MA 02474-5310
Phone: 781-641-2000
Fax: 781-641-2192

Amherst Alarm Inc.
435 Lawrence Bell Dr.
Amherst, NY 14221
Phone: 716-632-4600
Fax: 716-632-1156

Bay Alarm Co.
60 Berry Dr.
Pacheco, CA 94553
Phone: 925-935-1100
Fax: 925-808-4926

California Security Alarms, Inc.
1009 South Claremont St.
P.O. Box 5445
San Mateo, CA 94402
Phone: 650-570-6500
Fax: 650-574-0308

Central District Alarm, Inc.
6450 Clayton Ave.
St. Louis, MO 63139-3397
Phone: 314-647-2000
Fax: 314-647-7331

Central Station Monitoring
P.O. Box 1005
303 SW Zobrist
Estacada, OR 97023
Phone: 503-630-2565
Fax: 503-630-6630

Deep Valley Security
960 North State St.
Ukiah, CA 95482
Phone: 707-462-5200
Fax: 707-462-1478

Devcon Security Services Corp.
701 Park of Commerce Blvd., NW, Suite 200
Boca Raton, FL 33487
Phone: 561-998-1811
Fax: 561-998-1803

Doyle Security Systems
792 Calkins Rd.
Rochester, NY 14623
Phone: 585-461-6565
Fax: 585-442-9098

Electronic Security Corp. of America
254 Fairview Rd.
Woodlyn, PA 19094-1399
Phone: 610-833-5400
Toll Free: 1-800-224-3722
Fax: 610-521-5804

Electronix Systems Central Station Alarms, Inc.
1555 New York Ave.
Huntington Station, NY 11746-1707
Phone: 631-271-4000
Fax: 631-424-8510

EMERgency 24
4179 W. Irving Park Rd.
Chicago, IL 60641-2906
Phone: 773-777-0707
Toll Free: 1-800-827-3624
Fax: 773-286-1992

ESC Central, Inc.
3050 Guess Park Dr.
Birmingham, AL 35215
Phone: 205-520-0757
Toll Free: 1-800-268-3453
Fax: 205-520-5198

F. E. Moran, Inc. Alarm and Monitoring Services
2202 Fox Dr.
Champaign, IL 61820
Phone: 217-403-6444
Fax: 217-403-6442

Guardian Alarm Co. of Michigan, Inc.
Detroit Area Corporate Headquarters
20800 Southfield Rd.
Southfield, MI 48075
Phone: 248-423-1021
Fax: 248-395-1444

Guardian Protection Services
174 Thorn Hill Rd.
Warrendale, PA 15086-7528
Phone: 412-788-2580
Toll Free: 1-800-PROTECT
Fax: 412-787-2495

Holmes Electric Security Systems
127 Hay St.
Fayetteville, NC 28302
Phone: 910-483-6922
Toll Free: 1-800-426-9388
Fax: 910-483-1181

Home & Commercial Security, Inc.
44 Blanding Rd.
Rehoboth, MA 02769
Phone: 508-336-8833
Toll Free: 1-800-337-9469
Fax: 508-336-9049

Interface Security Systems, L.L.C.
3773 Corporate Center Dr.
Earth City, MO 63045
Phone: 314-595-0100
Fax: 314-595-0375

Jade Alarm Co.
7636 Troost Ave.
Kansas City, MO 64131
Phone: 816-333-5233
Fax: 816-444-2626

Lowitt Alarms & Security
25 Bethpage Rd.
Hicksville, NY 11801
Phone: 516-433-6960
Fax: 516-822-4490

Matson Alarm Co., Inc.
8401 N. Fresno St.
Fresno, CA 93720
Phone: 559-438-8000
Fax: 559-431-6291

Merchants Alarm Systems, Inc.
203 Paterson Ave.
Wallington, NJ 07057-1394
Phone: 973-779-4000
Fax: 973-779-7010

Monitronics International, Inc.
P.O. Box 814530
Dallas, TX 75381-4530
Phone: 972-243-7443
Fax: 972-243-1022

Nationwide Property Protection Services LLC
1 Nationwide Blvd.
Columbus, OH 43215
Phone: 614-249-7103
Fax: 614-249-3138

Pacific Alarm Systems, Inc.
4444 S. Sepulveda Blvd.
Culver City, CA 90230
Phone: 310-390-6661
Toll Free: 866-430-7233
Fax: 310-397-8796

Peak Alarm Co., Inc.
1534 S. Gladiola St.
Salt Lake City, UT 84104
Phone: 801-486-7231
Fax: 801-486-4931

Pro-Tec Monitoring Solutions Inc.
3950 12th Street NE
Calgary, Alberta T2E 8H9
Canada
Phone: 403-230-4175
Fax: 403-262-5888
Station: 403-277-6832

Protectron, Inc.
9120 Pascal-Gagnon
Montreal, Quebec H1P 2X4
Canada
Phone: 514-323-5000
Fax: 514-324-3780

Quick Response Monitoring Alarm Center
4734 Spring Rd.
Cleveland, OH 44131
Toll Free: 1-800-462-5353
Fax: 216-475-6527

Rapid Response Monitoring Services, Inc.
400 West Division St.
Syracuse, NY 13204
Toll Free: 1-800-932-3822
Fax: 315-422-8506

Response Center USA
11235 Gordon Rd., Suite 102
San Antonio, TX 78216
Phone: 866-489-4105

Safe Systems, Inc.
421 S. Pierce Ave.
Louisville, CO 80027
Phone: 303-444-1191
Fax: 303-449-0970

SAFECO Alarm Systems, Inc.
642 Broadway
Kingston, NY 12401
Phone: 845-338-4440;
845-338-0964

Safeguard—Home Security & Commercial Security Providers
16117 N. 76th St.
Scottsdale, AZ 85260
Phone: 480-609-6200
Fax: 480-609-6222

SDA Security
2054 State St.
P.O. Box 82567
San Diego, CA 92138-2567
Phone: 619-239-3473
Fax: 619-338-1209

Securall Monitoring Corp.
206 Washington Dr.
Brick, NJ 08724
Phone: 732-892-0700
Fax: 732-892-7916

Wayne Alarm Systems, Inc.
424 Essex St.
Lynn, MA 01902-3624
Phone: 781-595-0000
Fax: 781-215-5310

West Alarm Co., Inc.
2300 South Dakota Ave.
Sioux Falls, SD 57105
Phone: 605-339-1709
Toll Free: 1-800-303-1709
Fax: 605-334-8186

WH International Response Center
P.O. Box 330
6800 Electric Dr.
Rockford, MN 55373
Phone: 763-477-3144
Fax: 763-477-3153

WM Security Services
17340 Chanute Rd.
Houston, TX 77032
Phone: 281-784-4040
Fax: 281-784-4054

Registered Professional Locksmith Test

This test is based on the International Association of Home Safety and Security Professionals' Registered Security Professional Locksmith Program. If you earn a passing score, you should be able to pass other security certification and licensing examinations. To receive a Registered Locksmith Professional certificate, see the information after the test.

1. City codes often dictate the height and style of fences.

   a. True

   b. False

2. A hollow-core door is easy to break through.

   a. True

   b. False

3. In general, peepholes should be installed on windowless exterior doors.

   a. True

   b. False

4. A high-security strike box (or box strike) makes a door harder to kick in than a standard strike plate.

   a. True

   b. False

5. A skeleton key can be used to open warded bit-key locks.

   a. True

   b. False

6. If possible, house numbers should be visible from the street.

   a. True

   b. False

7. A standard electromagnetic lock includes a rectangular electromagnet and a rectangular wood and glass strike plate.

   a. True

   b. False

8. A blank is a key that fits two or more locks.

   a. True

   b. False

9. One difference between a bit key and barrel key is the bit key has a hollow shank.

   a. True

   b. False

10. A key-in-knob lock typically is used to secure windows.

    a. True

    b. False

11. The Egyptians are credited with inventing the first lock to be based on the locking principle of today's pin-tumbler lock.

    a. True

    b. False

12. A jimmy-proof deadlock typically is the most secure type of lock for sliding glass doors.

    a. True

    b. False

13. Lock picking is the most common way homes are burglarized.

    a. True

    b. False

14. Lock impressioning is the most common way homes are burglarized.

    a. True

    b. False

15. A long-reach tool and wedge are commonly used to open locked automobiles.

    a. True

    b. False

16. It is legal for locksmiths to duplicate a U.S. Post Office box key at the request of the box renter—if the box renter shows a current passport or driver's license.

    a. True

    b. False

17. The Romans are credited with inventing the pin-tumbler lock.

    a. True

    b. False

18. Five common keyway groove shapes are left angle, right angle, square, *V*, and round.

    a. True

    b. False

19. To pick open a pin-tumbler cylinder, you usually need a pick and a torque wrench.

    a. True

    b. False

20. A Door Reinforcer makes a lock harder to pick open.

    a. True

    b. False

21. Common door-lock backsets include

    a. 2-1/2 inch and 2-3/8 inch.

    b. 3 inch and 2-3/4 inch.

    c. 3/4 inch and 2-7/8 inch.

    d. 2-3/8 inch and 2-3/4 inch.

22. How many sets of pin tumblers are in a typical pin-tumbler house door lock?

    a. 3 or 4

    b. 5 or 6

    c. 11 or 12

    d. 1 or 8

23. Which lock is unpickable?

    a. A Medeco biaxial deadbolt

    b. A Grade 2 Titan

    c. The Club steering wheel lock

    d. None of the above

24. Which are basic parts of a standard key-cutting machine?

    a. A pair of vises, a key stop, and a grinding stylus

    b. Two cutter wheels, a pair of vises, and a key shaper

    c. A pair of vises, a key stylus, and a cutter wheel

    d. A pair of styluses, a cutter wheel, and a key shaper

25. What are two critical dimensions for code-cutting cylinder keys?

    a. Spacing and depth

    b. Bow size and blade thickness

    c. Blade width and keyhole radius

    d. Shoulder width and bow size

26. Which manufacturer is best known for its low-cost residential key-in-knob locks?

    a. Kwikset Corporation

    b. Medeco Security Locks

    c. The Key-in-Knob Corporation

    d. ASSA

27. The most popular mechanical lock brands in the United States include

    a. Yale, Master, Corby, and Gardall

    b. Yale, Kwikset, Master, and TuffLock

    c. Master, Weiser, Kwikset, and Schlage

    d. Master, Corby, Gardall, and Tufflock

28. A mechanical lock that is operated mainly by a pin-tumbler cylinder is commonly called

    a. A disk-tumbler pinned lock

    b. A cylinder-pin lock

    c. A mechanical cylinder-pin lock

    d. A pin-tumbler cylinder lock

29. Burglars target garage doors because

    a. People keep property that is easy to fence in garages.

    b. A garage attached to the house provides a discreet way to break into the house.

    c. A garage door with thin or loose panels can be accessed without opening the door.

    d. All the above.

30. Glass is a deterrent to burglars because

    a. It slows down a burglar.

    b. Broken shards of glass can injure a burglar.

    c. Shattering glass is noisy and attracts attention.

    d. All of the above.

31. If you don't feel secure about glass in a window, you can increase security by

    a. Replacing the glass with carbonated glass or antihammer plastic.

    b. Replacing the glass with impact-resistant acrylic or polycarbonate, or with high-security glass.

    c. Covering the glass with bullet-proof paint.

    d. All the above.

32. Warded bit-key locks

    a. Provide high security.

    b. Provide little security.

    c. Are hard to open without the right key.

    d. Are usually the best choice for use on an exterior door.

33. Which of the following key combinations provides the most security?

    a. 55555

    b. 33333

    c. 243535

    d. 35353

34. Which of the following key combinations provides the least security?

    a. 243535

    b. 1111

    c. 321231

    d. 22222

35. A blank is basically just

    a. A change key with cuts on one side only.

    b. An uncut or uncombinated key.

    c. Any key with no words or numbers on the bow.

    d. A master key with no words or numbers on the bow.

36. You often can determine the number of pin stacks or tumblers in a cylinder by

    a. Its key-blade length.

    b. Its key-blade thickness.

    c. The key-blank manufacturer's name on the bow.

    d. The material of the key.

37. Spool and mushroom pins

    a. Make keys easier to duplicate.

    b. Can hinder normal picking attempts.

    c. Make a lock easier to pick.

    d. Make keys harder to duplicate.

38. As a general rule, the 10-cut wafer sidebar locks by General Motors have

    a. A sum total of cut depths that must equal an even number.

    b. Up to four of the same depth cut in the 7, 8, 9, and 10 spaces.

    c. A maximum of five number 1 depths in a code combination.

    d. At least one 4–1 or 1–4 adjacent cuts.

39. When drilling open a standard pin-tumbler cylinder, position the drill bit

    a. At the first letter of the cylinder.

    b. At the shear line, in alignment with the top and bottom pins.

    c. Directly below the bottom pins.

    d. Directly above the top pins.

40. When viewed from the exterior side, a door that opens inward and has hinges on the right side is

    a. A left-hand door

    b. A right-hand door

    c. A left-hand reverse-bevel door

    d. A right-hand reverse-bevel door

41. A utility patent

    a. Relates to a product's appearance, is granted for 14 years, and is renewable.

    b. Relates to a product's function, is granted for 17 years, and is nonrenewable.

    c. Relates to a product's appearance, is granted for 17 years, and is renewable.

    d. Relates to a product's function, is granted for 35 years, and is nonrenewable.

42. To earn a UL-437 rating, a sample lock must

    a. Pass a performance test.

    b. Use a patented key.

    c. Use hardened-steel mounting screws and mushroom and spool pins.

    d. Pass an attack test using common hand and electric tools such as drills, saw blades, puller mechanisms, and picking tools.

43. Tumblers are

    a. Small metal objects that protrude from a lock's cam to operate the bolt.

    b. Fixed projections on a lock's case.

    c. Small pins, usually made of metal, that move within a lock's case to prevent unauthorized keys from entering the keyhole.

    d. Small objects, usually made of metal, that move within a lock cylinder in ways that obstruct a lock's operation until an authorized key or combination moves them into alignment.

44. Electric switch locks

    a. Are mechanical locks that have been modified to operate with battery power.

    b. Complete and break an electric current when an authorized key is inserted and turned.

    c. Are installed in metal doors to give electric shocks to intruders.

    d. Are mechanical locks that have been modified to operate with alternating current (ac) electricity instead of with a key.

45. A popular type of lock used on GM cars is

    a. A Medeco pin tumbler

    b. An automotive bit key

    c. A sidebar wafer

    d. An automotive tubular key

46. When cutting a lever-tumbler key by hand, the first cut should be the

    a. Lever cut

    b. Stop cut

    c. Throat cut

    d. Tip cut

47. How many possible key changes does a typical disk-tumbler lock have?

    a. 1,500

    b. 125

    c. A trillion

    d. 25

48. Which manufacturer is best known for its interchangeable core locks?

    a. Best Lock

    b. Kwikset Corporation

    c. ILCO Interchangeable Core Corporation

    d. Interchangeable Core Corporation

49. James Sargent is famous for

    a. Inventing the Sargent key-in-knob lock.

    b. Inventing the time lock for banks.

    c. Inventing the double-acting lever-tumbler lock.

    d. Being the first person to pick open a Medeco biaxial cylinder.

50. Which are common parts of a combination padlock?

    a. Shackle, case, bolt

    b. Spacer washer, top pins, cylinder housing

    c. Back cover plate, case, bottom pins

    d. Wheel pack base plate, wheel pack spring, top and bottom pins

51. General Motors' ignition lock codes generally can be found
    a. On the ignition lock
    b. On the passenger-side door
    c. Below the Vehicle Identification Number (VIN) on the vehicle's engine
    d. Under the vehicle's brake pedal

52. Which code series is used commonly on Chrysler door and ignition locks?
    a. EP 1-3000
    b. CHR 1-5000
    c. CRY 1-4000
    d. GM 001-6000

53. How many styles of lock pawls does General Motors use in its various car lines?
    a. One
    b. Five
    c. More than 5
    d. Three

54. The double-sided (or 10-cut) Ford key
    a. Has five cuts on each side; one side operates the trunk and door, whereas the other side operates the ignition.
    b. Has five cuts on each side; either side can operate all locks of a car.
    c. Has ten cuts on each side; one side operates the trunk and door, whereas the other side operates only the ignition.
    d. Has ten cuts on each side.

55. Usually the simplest way to change the combination of a double-bitted cam lock is to
    a. Rearrange the positions of two or more tumblers.
    b. Remove two tumblers and replace them with new tumblers.
    c. Remove the tumbler assembly and replace it with a new one.
    d. Connect a new tumbler assembly to the existing one.

56. When shimming a pin-tumbler cylinder open

    a. Use the key to insert the shim into the keyway.

    b. Insert the shim into the keyway without the key.

    c. Insert the shim along the left side of the cylinder housing.

    d. Insert the shim between the plug and cylinder housing between the top and bottom pins.

57. A lock is any

    a. Barrier or closure that restricts entry.

    b. Fastening device that allows a person to open and close a door, window, cabinet, drawer, or gate.

    c. Device that incorporates a bolt, cam, shackle, or switch to secure an object—such as a door, drawer, or machine—to a closed, locked, on, or off position and that provides a restricted means—such as a key or combination—of releasing the object from that position.

    d. Device or object that restricts entry to a given premise.

58. Which wheel in a safe lock is closest to the dial?

    a. Wheel 1

    b. Wheel 2

    c. Wheel 3

    d. Wheel 0

59. Which type of safe is best for protecting paper documents from fire?

    a. Paper safe

    b. Money safe

    c. UL-listed record safe

    d. Patented burglary safe

60. Which type of cylinder is typically found on an interlocking deadbolt (or jimmy-proof deadlock)?

    a. Bit-key cylinder

    b. Key-in-knob cylinder

    c. Rim cylinder

    d. Tubular deadbolt cylinder

Registered Professional Locksmith Answer Sheet

Make a photocopy of this answer sheet and mark your answers
on it.

1. ____	21. ____	41. ____
2. ____	22. ____	42. ____
3. ____	23. ____	43. ____
4. ____	24. ____	44. ____
5. ____	26. ____	45. ____
6. ____	27. ____	46. ____
7. ____	28. ____	47. ____
8. ____	29. ____	48. ____
9. ____	30. ____	49. ____
10. ____	31. ____	50. ____
11. ____	32. ____	51. ____
12. ____	32. ____	52. ____
13. ____	33. ____	53. ____
14. ____	34. ____	54. ____
15. ____	35. ____	55. ____
16. ____	36. ____	56. ____
17. ____	37. ____	57. ____
18. ____	38. ____	58. ____
19. ____	39. ____	59. ____
20. ____	40. ____	60. ____

To receive your Registered Professional Locksmith certificate, just submit your answers to this test (a passing score is
70 percent) and enclose a check or money order for US$50
(nonrefundable) payable to IAHSSP. If you join the IAHSSP
(see Appendix G), there is no fee to take the RPL test. Also
enclose copies of any two of the following items (do not send
original documents because they will not be returned):

- City or state locksmith license
- Driver's license or state-issued identification
- Locksmith suppliers' invoice
- Certificate from locksmithing or security school or program
- Yellow Pages listing
- Business card or letterhead from your company
- Association membership card or certificate

- Locksmithing, safe technician, or alarm installer bond card or certificate
- Letter from your employer or supervisor on company letterhead stating that you work as a locksmith or security professional
- Letter of recommendation from a Registered Professional Locksmith or Registered Security Professional
- Copy of an article you've had published in a locksmithing or security trade journal
- ISBN number and title of locksmith- or security-related book you wrote

   Send everything to IAHSSP, Box 2044, Erie, PA 16512-2044. Please allow six to eight weeks for processing.

Name: _____

Title: _____

Company Name: _____

Address: _____

City/State/ZIP: _____

Telephone Number: _____

E-mail Address: _____

International
Association of
Home Safety and
Security Professionals, Inc.

Membership Application

Membership in the International Association of Home Safety and Security Professionals, Inc. (or "IAHSSP") is open to security consultants, alarm-systems installers, locksmiths, home automation system installers, safe technicians, law enforcement officers, educators, manufacturers, and others who provide safety or security-related products or services.

To become a member, complete this application and return it along with the $25 application fee, your annual membership fee, and any materials that may help determine your eligibility for membership—such as a resume, copies of news clips, letters of recommendation, copies of membership or certification certificates, copy of "yellow pages" listing, clips of articles you've written, etc. The more documentation you send, the better. Don't send originals, because none of the material you send will be returned to you.

Your membership application will be approved or denied within 60 days after it is received. If denied, your membership fee will be promptly refunded. (The application fee is nonreturnable.)

Please **print or type** your answers to the following questions:

Circle the appropriate title: Mr.   Mrs.   Ms.   Dr.

First Name _____ MI _____ Last Name _____

Mailing Address _____

City/State/Zip _____

Home address (if different from mailing) _____

City/State/Zip _____

Telephone (____) _____ Fax (____) _____

Birthdate: _____ Social Security # _____

Business Name _____ Employer I.D. (Corporations) _____

Business Address _____ City/State/Zip _____

Business Phone (____) _____ Fax (____) _____

Are you a (check one only):   ☐ sole owner   ☐ partner   ☐ student   ☐ corporate officer

☐ employee   ☐ apprentice   ☐ instructor   ☐ researcher

If employee, give name of owner or your supervisor _____

If student, give name/address of school _____

Are you a safety or security professional? _____ For how long? _____

What type of safety or security-related work do you do? _____

How did you learn your trade or profession? _____

Which type of membership are you applying for (check one):

Professional — for persons 18 years or older who have been safety or security professionals for at least five years.

Associate — for students, apprentices, authors, and others who have
for at least five years.

Allied — for manufacturers, distributors, schools, publishers, and other businesses and institutions that have been **providing services** or products to safety or security professionals for at least five years.

I certify that all statements herein are true. If accepted as a member, I agree to abide by all rules and regulations of IAHSSP. I understand that only professional and allied members in good standing may reproduce the organization's logo, and that it may be used by such members only in their "yellow pages" listings, on their stationery, and in their advertisements for the purpose of identifying them as members of IAHSSP. I will not use the IAHSSP logo in any illegal, misleading, or distasteful manner, nor will I direct any agent, employee, employer, or representative to use the logo in such a manner. I, the signer of this application, and my company listed herein, jointly and severally agree to indemnify IAHSSP against any cost, expenses, or damages arising from or appurtenant to my or my companys' direct or indirect misuse of the logo. I understand the IAHSSP may at any time, with or without cause, revoke or restrict any right I or my company may have to use the logo.

Signature _____ Title _____ Date _____

Annual Dues: Professional Member $100. Allied Member $200. Associate Member $75.

Checklist of Enclosures: ☐ $25 application fee   ☐ Annual Membership dues

$ _____   ☐ Resume   ☐ Letterhead   ☐ Business card   ☐ Telephone book listing

☐ Advertisement   ☐ License   ☐ Membership certificate   ☐ Certification certificate

☐ Other _____

Make Check or Money Order Payable to: IAHSSP (U.S. funds only.)
Return payment, application, and all attachments to:
        IAHSSP
        Box 2044
        Erie, PA 16512-2004

For Office Use Only:

Date Received _____ Payment Enclosed $ _____ Form _____

Date Approved/Denied _____ Comments _____

Glossary

The Professional Glossary of Terms Relating to Cylinders, Keys, and Master Keying. By the Master Keying Study Group of the AOLA-Sponsored National Task Group for Certified Training Programs. Courtesy of the Lock Industry Standards and Training (LIST) Council.

A, AA, AA1, 1AA, and so forth
> *See* "key symbol." 1. *See* "keying symbol." 2. *See* "standard key coding system."

Across-the-key-progression
> (n.) *See* "total position progression."

Actuator
> (n.) A device, usually connected to a cylinder, which, when activated, may cause a lock mechanism to operate.

Adjacent cut differential
> (n.) *See* "maximum adjacent-cut specification."

Adjustable mortise cylinder
> (n.) Any mortise cylinder whose length can be adjusted for a better fit in doors of varying thickness.

AFTE
> (abb.) Association of Firearm and Toolmark Examiners.

AHC
> (abb.) Architectural Hardware Consultant (as certified by DHI).

All-section key blank
> (n.) The key section that enters all keyways of a multiplex key system.

ALOA
> (abb.) Associated Locksmiths of America.

Angle of cut
> (n.) 1. *See* "cut angle." 2. *See* "degree of rotation."

Angularly bitted key
> (n.) A key that has cuts made into the blade at various degrees of rotation from the perpendicular.

ANSI

> (abb.) American National Standards Institute.

Armored front

> (n.) *See* "face plate."

ASIS

> (abb.) American Society for Industrial Security.

Associated change key

> (n.) A change key that is related directly to particular master key(s) through the use of constant cuts.

Associated master key

> (n.) A master key that has particular change keys related directly to its combination through the use of constant cuts.

ASTM

> (abb.) American Society for Testing and Materials.

Back of blade

> (n.) *See* "bottom of blade."

Back plate

> (n.) A thin piece of metal, usually with a concave portion, used with machine screws to fasten certain types of cylinders to a door.

Backed off blade

> (n.) *See* "radiused blade bottom."

Ball bearing

> (n.) 1. A metal ball used in the pin stack to accomplish some types of hotel or construction keying. 2. A ball, usually made of steel, used by some lock manufacturers as the bottom element in the pin stack in one or more pin chambers. 3. Any metal ball used as a tumbler's primary component.

Ball end pin

> (n.) *See* "bottom pin."

Barrel

> (n.) *See* "cylinder plug."

Bell type key

> (n.) A key whose cuts are in the form of wavy grooves milled into the flat sides of the key blade. The grooves usually run the entire length of the blade.

BHMA

> (abb.) Builders Hardware Manufacturers Association.

Bible

(n.) The portion of the cylinder shell that houses the pin chambers, especially those of a key-in-knob cylinder or certain rim cylinders.

Bicentric cylinder

(n.) A cylinder that has two independent plugs, usually with different keyways. Both plugs are operable from the same face of the cylinder. It is designed for use in extensive master key systems.

Bidirectional cylinder

(n.) A cylinder that may be operated in a clockwise and counterclockwise direction by a single key.

Binary cut key

(n.) A key whose combination only allows for two possibilities in each bitting position: cut/no cut.

Binary type cylinder or lock

(n.) A cylinder or lock whose combination only allows for two bitting possibilities in each bitting position.

Bit

(n.) 1. The part of the key that serves as the blade, usually for use in a warded or lever tumbler lock. 2. (n.) *See* "key cut(s)." 3. (v.) To cut a key.

Bitting

(n.) 1. The number(s) that represent(s) the dimensions of the key cut(s). 2. The actual cut(s) or combination of a key.

Bitting depth

(n.) The depth of a cut that is made into the blade of a key.

Bitting list

(n.) A listing of all the key combinations used within a system. The combinations are usually arranged in order of the blind code, direct code, and/or key symbol.

Bitting position

(n.) 1. The location of a key cut. 2. *See* "spacing."

Blade

(n.) The portion of a key that may contain the cuts and/or milling.

Blade tumbler

(n.) *See* "disc tumbler."

Blank

(n.) 1. *See* "key blank." (adj.) 2. uncut.

Blind code

(n.) A designation, unrelated to the bitting, assigned to a particular key combination for future reference when additional keys or cylinders may be needed.

Block master key

(n.) The one pin master key for all combinations listed as a block in the standard progression format.

Blocking ring

(n.) *See* "cylinder collar."

Blockout key

(n.) *See* "lockout key."

Bottom of blade

(n.) The portion of the blade opposite the cut edge of a single bitted key.

Bottom pin

(n.) Usually, a cylindrically shaped tumbler that may be conical, ball shaped, or chisel pointed on the end that makes contact with the key.

Bow

(n.) The portion of the key that serves as a grip or handle.

Bow stop

(n.) A type of stop located near the key bow.

Broach

(n.) 1. A tool used to cut the keyway into the cylinder plug. (v.) 2. To cut the keyway into a cylinder plug with a broach.

Builders' master key

(n.) *See* "construction master key."

Building master key

(n.) A master key that operates all or most master keyed locks in a given building.

Bypass key

(n.) The key that operates a key override cylinder.

Cam

(n.) A lock or cylinder component which transfers the rotational motion of a key or cylinder plug to the bolt works of a lock 2. The bolt of a cam lock.

Cam lock

(n.) A complete locking assembly in the form of a cylinder whose cam is the actual locking bolt.

Cap

(n.) 1. A spring cover for a single pin chamber. (n.) 2. A part that may serve as a plug retainer and/or a holder for the tail-piece. (v.) 3. To install a cap.

Capping block

(n.) A holding fixture for certain interchangeable cores that aids in the installation of the caps.

Cell

(n.) *See* "pin chamber."

Central key system

(n.) *See* "maison key system."

Chain key system

(n.) *See* "selective key system."

Chamber

(n.) Any cavity in a cylinder plug and/or shell that houses the tumbler(s).

Change key

(n.) 1. A key that operates only one cylinder or one group of keyed alike cylinders in a keying system. 2. *See* "reset key."

Change key constant

(n.) *See* "constant cut."

Change key section

(n.) *See* "single key section."

Changeable bit key

(n.) A key that can be recombined by exchanging and/or rearranging portions of its bit or blade.

Chart

(n.) 1. *See* "pinning chart." 2. *See* "progression list." 3. *See* "bitting list." 4. *See* "key system schematic."

Chip

(n.) *See* "master pin."

CK

(abb.) 1. change key. (abb.) 2. control key.

Clutch

(n.) That part of the profile cylinder that transfers rotational motion from the inside or outside element to a common cam or actuator.

CMK

(abb.) Construction master key.

CMK'd

(abb.) Construction master keyed.

Code

(n.) 1. A designation assigned to a particular key combination for reference when additional keys or cylinders may be needed. *See also* "blind code," "direct code," and "key symbol." (v.) 2. *See* "combinate."

Code key

(n.) A key cut to a specific code rather than duplicated from a pattern key. It may or may not conform to the lock manufacturer's specifications.

Code list

(n.) 1. *See* "bitting list." 2. *See* "progression list."

Code machine

(n.) *See* "key coding machine."

Code number

(n.) 1. *See* "blind code." 2. *See* "direct code."

Code original key

(n.) A code key that conforms to the lock manufacturer's specifications.

Column master key

(n.) *See* "vertical group master key."

Combinate

(v.) To set a combination in a lock, cylinder, or key.

Combination

(n.) The group of numbers that represent the bitting of a key and/or the tumblers of a lock or cylinder.

Combination wafer

(n.) A type of disc tumbler used in certain binary type disc tumbler key-in-knob locks. Its presence requires that a cut be made in that position of the operating key(s).

Common keyed

(adj.) *See* "maison key system."

Compensate drivers

(v.) 1. To select longer or shorter top pins, depending on the length of the pin stack, to achieve a uniform pin stack height. 2. *See* "graduated drivers."

Complementary keyway

> (n.) Usually, a disc tumbler keyway used in master keying. It accepts keys of different sections whose blades contact different bearing surfaces of the tumblers.

Composite keyway

> (n.) A keyway that has been enlarged to accept more than one key section, often key sections of more than one manufacturer.

Concealed shell cylinder

> (n.) A specially constructed (usually mortise) cylinder. Only the plug face is visible when the lock trim is in place.

Connecting bar

> (n.) *See* "tailpiece."

Constant cut

> (n.) 1. Any bitting(s) that are identical in corresponding positions from one key to another in a keying system. They usually serve to group these keys together within a given level of keying, and/or link them with keys of other levels. 2. *See* "rotating constant."

Construction breakout key

> (n.) A key used by some manufacturers to render all construction master keys permanently inoperative.

Construction core

> (n.) An interchangeable or removable core designed for use during the construction phase of a building. The cores are normally keyed alike and, on completion of construction, they are to be replaced by the permanent system's cores.

Construction master key

> (n.) A key normally used by construction personnel for a temporary period during building construction. It may be rendered permanently inoperative without disassembling the cylinder.

Construction master keyed

> (adj.) Of, or pertaining to, a cylinder that is or is to be operated temporarily by a construction master key.

Control cut

> (n.) 1. Any bitting that operates the retaining device of an interchangeable or removable core. 2. *See* "constant cut."

Control key

> (n.) 1. A key whose only purpose is to remove and/or install an interchangeable or removable core. 2. A bypass key used to operate and/or reset some combination type locks. 3. A key that allows disassembly of some removable cylinder locks.

Control lug
>   (n.) That part of an interchangeable or removable core retaining device that locks the core into its housing.

Control sleeve
>   (n.) The part of an interchangeable core retaining device that surrounds the plug.

Controlled cross keying
>   (n.) A condition in which two or more different keys of the same level of keying and under the same higher level key(s) operate one cylinder by design. For example, XAA1 operated by AA2 (but not XAA1 operated by AB1). NOTE: This condition could severely limit the security of the cylinder and the maximum expansion of the system when (1) more than a few of these different keys operate a cylinder, or (2) more than a few differently cross keyed cylinders per system are required.

Core
>   (n.) A complete unit, often with a Figure 8 shape that usually consists of the plug, shell, tumblers, springs, plug retainer, and spring cover(s). This is primarily used in removable and interchangeable core cylinders and locks.

CPP
>   (abb.) Certified Protection Professional (as certified by ASIS).

Cross keyed
>   (adj.) *See* "cross keying."

Cross keying
>   (n.) The deliberate process of combinating a cylinder (usually in a master key system) to two or more different keys that would not normally be expected to operate it together. *See also* "controlled cross keying," and "uncontrolled cross keying."

CSI
>   (abb.) Construction Specifiers Institute.

Cut
>   (n.) 1. *See* "key cut(s)." (v.) 2. To make cuts into a key blade.

Cut angle
>   (n.) 1. A measurement, usually expressed in degrees, for the angle between the two sides of a key cut. 2. *See* "degree of rotation."

Cut depth
>   (n.) 1. *See* "bitting depth." 2. *See* "root depth."

Cut edge
>   (n.) The portion of the key blade that contains the cuts.

Cut key

    (n.) A key that has been bitted or combinated.

Cut profile

    (n.) *See* "key cut profile."

Cut root

    (n.) The bottom of a key cut.

Cut root shape

    (n.) The shape of the bottom of a key cut. It may have a flat or radius of a specific dimension or be a perfect V.

Cut rotation

    (n.) *See* "degree of rotation."

Cutter

    (n.) The part of a key machine that makes the cuts into the key blank.

Cylinder

    (n.) A complete operating unit that usually consists of the plug, shell, tumblers, springs, plug retainer, a cam/tailpiece or other actuating device, and all other necessary operating parts.

Cylinder assembly

    (n.) *See* "cylinder."

Cylinder bar

    (n.) *See* "tailpiece."

Cylinder blank

    (n.) A dummy cylinder that has a solid face and no operating parts.

Cylinder clip

    (n.) A spring steel device used to secure some types of cylinders.

Cylinder collar

    (n.) A plate or ring installed under the head of a cylinder to improve appearance and/or security.

Cylinder guard

    (n.) A protective cylinder mounting device.

Cylinder key

    (n.) A broad generic term including virtually all pin and disc tumbler keys.

Cylinder plug

    (n.) *See* "plug."

Cylinder ring

    (n.) *See* "cylinder collar."

Cylinder rose

> (n.) *See* "cylinder collar."

Cylinder shell

> (n.) *See* "shell."

Day key

> (n.) *See* "change key."

Declining step key

> (n.) A key whose cuts are progressively deeper from bow-to-tip.

Decode

> (v.) To determine a key combination by physical measurement of a key and/or cylinder parts.

Degree of rotation

> (n.) A specification for the angle at which a cut is made into a key blade as referenced from the perpendicular; e.g., right (R or 2), left (L or 1) or center (= perpendicular) (C). This specification is typically used for some high security keys.

Department master key

> (n.) A master key that operates all or most master keyed locks of a given department.

Depth

> (n.) 1. *See* "bitting depth." 2. *See* "root depth."

Depth key set

> (n.) A set of keys used to make a code original key on a key duplicating machine to a lock manufacturer's given set of key bitting specifications. Each key is cut with the correct spacing to one depth only in all bitting positions, with one key for each depth.

Derived series

> (n.) A series of blind codes and bittings that are directly related to those of another bitting list.

Detainer disc

> (n.) *See* "rotary tumbler."

DHI

> (abb.) Door and Hardware Institute.

Dimple

> (n.) A key cut in a dimple key.

Dimple key

> (n.) A key whose cuts are drilled or milled into its blade surfaces. The cuts normally do not change the blade silhouette.

Direct cods

(n.) A designation assigned to a particular key that includes the actual combination of the key.

Disc

(n.) 1. *See* "disc tumbler." 2. *See* "master pin." 3. *See* "rotary tumbler."

Disc tumbler

(n.) 1. A flat tumbler that must be drawn into the cylinder plug by the proper key, so none of its extremities extends into the shell. 2. A flat, usually a rectangular tumbler with a gate that must be aligned with a sidebar by the proper key.

Display key

(n.) A special change key in a hotel master key system that allows access to one designated guest room, even if the lock is in the shut-out mode. It may also act as a shut-out key for that room.

Double bitted key

(n.) A key bitted on two opposite surfaces.

Double pin

(v.) To place more than one master pin in a single pin chamber.

Double-sided key

(n.) *See* "double bitted key."

Driver

(n.) *See* "top pin."

Driver spring

(n.) A spring placed on top of the pin stack to exert pressure on the pin tumblers.

Drop

(n.) 1. *See* "increment." 2. A pivoting or swinging dust cover.

Dummy cylinder

(n.) A nonfunctional facsimile of a rim or mortise cylinder used for appearance only, usually to conceal a cylinder hole.

Duplicate

1. (n.) *See* "duplicate key." 2. (v.) To copy.

Duplicate blank

(n.) *See* "nonoriginal key blank."

Duplicate key

(n.) Any key reproduced from a pattern key.

Dust cover

(n.) A device designed to prevent foreign matter from entering a mechanism through the keyway.

Dustproof cylinder

(n.) A cylinder designed to prevent foreign matter from entering either end of the keyway.

Effective plug diameter

(n.) The dimension obtained by adding the root depth of a key cut to the length of its corresponding bottom pin, which establishes a perfect shear line. This is not necessarily the same as the *actual* plug diameter.

Ejector hole

(n.) A hole found on the bottom of certain interchangeable cores under each pin chamber. It provides a path for the ejector pin.

Ejector pin

(n.) A tool used to drive all the elements of a pin chamber out of certain interchangeable cores.

Emergency key

(n.) 1. *See* "emergency master key." 2. The key that operates a privacy function lockset.

Emergency master key

(n.) A special master key that usually operates all guest room locks in a hotel master key system at all times, even in the shut-out mode. This key may also act as a shut-out key.

EMK

(abb.) Emergency master key.

Encode

(v.) *See* "combinate."

ENG

Symbol for engineer's key.

Engineer's key

(n.) A selective master key used by maintenance personnel to operate many locks under different master keys in a system of three or more levels of keying.

Escutcheon

(n.) A surface-mounted trim that enhances the appearance and/or security of a lock installation.

Extractor key

(n.) A tool that normally removes a portion of a two-piece key or blocking device from a keyway.

Face plate

(n.) 1. A mortise lock cover plate exposed in the edge of the door. 2. *See* "scalp."

Factory original key

(n.) The cut key furnished by the lock manufacturer for a lock or cylinder.

False plug

(n.) *See* "plug follower."

Fence

(n.) 1. A projection on a lock bolt that prevents movement of the bolt unless it can enter gates of properly aligned tumblers. 2. *See* "sidebar."

File key

(n.) *See* "pattern key."

Finish

(n.) A material, coloring, and/or texturing specification.

Fireman's key

(n.) A key used to override normal operation of elevators, bringing them to the ground floor.

First generation duplicate

(n.) A key that was duplicated using a factory original key or a code original key as a pattern.

First key

(n.) Any key produced without the use of a pattern key.

Five-column progression

(n.) A process wherein key bittings are obtained by using the cut possibilities in five columns of the key bitting array.

Five-pin master key

(n.) A master key for all combinations obtained by progressing five bitting positions.

Flexible head mortise cylinder

(n.) An adjustable mortise cylinder that can be extended against spring pressure to a slightly longer length.

Floating master key

(n.) 1. *See* "unassociated master key." 2. *See* "selective master key."

Floor master key

(n.) A master key that operates all or most master keyed locks on a particular floor of a building.

Follower

    (n.) *See* "plug follower."

Formula

    (n.) *See* "key bitting array."

Four-column progression

    (n.) A process wherein key bittings are obtained by using the cut possibilities in four columns of the key bitting array.

Four-pin master key

    (n.) A master key for all combinations obtained by progressing four bitting positions.

Gate

    (n.) A notch cut into the edge of a tumbler to accept a fence or sidebar.

Gauge key

    (n.) 1. *See* "depth key set." 2. *See* "set-up key."

Genuine key blank

    (n.) *See* "original key blank."

GGGMK

    (abb.) Great-great grand master key.

GGGMKd

    (abb.) Great-great grand master keyed.

GGM

    (abb.) Great grand master key.

GGMK

    (abb.) Great grand master key.

GGMK'd

    (abb.) Great grand master keyed.

Ghost key

    (n.) *See* "incidental master key."

GM

    (abb.) Grand master key.

GMK

    (abb.) Grand master key.

GMK section

    (abb.) Grand master key section.

GMK'd

    (abb.) Grand master keyed.

Graduated drivers

    (n.) 1. A set of top pins of different lengths. Usage is based on the height of the rest of the pin stack, to achieve a uniform pin stack height. 2. *See* "compensate drivers."

Grand master key

    (n.) The key that operates two or more separate groups of locks, each of which is operated by a different master key.

Grand master key section

    (n.) 1. *See* "multisection key blank." 2. *See* "all-section key blank."

Grand master key system

    (n.) A master key system that has exactly three levels of keying.

Grand master keyed

    (adj.) Of, or pertaining to, a lock or cylinder that is or is to be keyed into a grand master key system.

Great grand master key

    (n.) The key that operates two or more separate groups of locks that are operated by a different grand master key.

Great grand master key system

    (n.) A master key system that has exactly four levels of keying.

Great grand master keyed

    (adj.) Of, or pertaining to, a lock or cylinder that is or is to be keyed into a great grand master key system.

Great-great grand master key

    (n.) The key that operates two or more separate groups of locks, each of which is operated by different great grand master keys.

Great-great grand master key system

    (n.) A master key system that has five or more levels of keying.

Great-great grand master keyed

    (adj.) Of, or pertaining to, a lock or cylinder that is or is to be keyed into a great-great grand master key system.

Grooving

    (n.) *See* "key milling."

Guard key

    (n.) A key that must be used in conjunction with a renter's key to unlock a safe deposit lock. This key is usually the same for every lock within an installation.

Guest key

    (n.) A key in a hotel master key system that is normally used to unlock only the one guest room for which it was intended, but it does not operate the lock in the shut-out mode.

Guide

    (n.) That part of a key machine that follows the cuts of a pattern key or template during duplication.

Guide keys

    (n.) *See* "depth key set."

Hardware schedule

    (n.) A listing of the door hardware used on a particular job. This includes the types of hardware, manufacturers, locations, finishes, and sizes. It should include a keying schedule specifying how each locking device is to be keyed.

HGM

    (abb.) Horizontal group master key.

High-security cylinder

    (n.) A cylinder that offers a greater degree of resistance to any or all of the following: picking, impressioning, key duplication, drilling, or other forms of forcible entry.

High-security key

    (n.) A key for a high-security cylinder.

HKP

    (abb.) Housekeeper's key.

Hold and vary

    (n.) *See* "rotating constant method."

Hold open cylinder

    (n.) A cylinder provided with a special cam that holds a latch bolt in the retracted position when so set by the key.

Holding fixture

    (n.) A device that holds cylinder plugs, cylinders, housings, and/or cores to facilitate the installation of tumblers, springs, and/or spring covers.

Hollow driver

    (n.) A top pin hollowed out on one end to receive the spring, typically used in cylinders with extremely limited clearance in the pin chambers.

Horizontal group master key

    (n.) The two-pin master key for all combinations listed in all blocks in a line across the page in the standard progression format.

Housekeeper's key

(n.) A selective master key in a hotel master key system that may operate all guest and linen rooms, and other housekeeping areas.

Housing

(n.) The part of a locking device that is designed to hold a core.

Imitation blank

(n.) *See* "nonoriginal key blank."

Impression

(n.) 1. The mark made by a tumbler on its key cut. (v.) 2. To fit a key by the impression technique.

Impression technique

(n.) A means of fitting a key directly to a locked cylinder by manipulating a blank in the keyway and cutting the blank where the tumblers have made marks.

Incidental master key

(n.) A key cut to an unplanned shear line created when the cylinder is combinated to the top master key and a change key.

Increment

(n.) A usually uniform increase or decrease in the successive depths of a key cut that must be matched by a corresponding change in the tumblers.

Indicator

(n.) A device that provides visual evidence that a deadbolt is extended or a lock is in the shut-out mode.

Indirect code

(n.) *See* "blind code."

Individual key

(n.) 1. An operating key for a lock or cylinder that is not part of a keying system. 2. *See* "change key."

Interchange

(n.) *See* "key interchange."

Interchangeable core

(n.) A key removable core that can be used in all or most of the core manufacturer's product line. No tools (other than the control key) are required for removal of the core.

Interlocking pin tumbler

(n.) A type of pin tumbler that is designed to be linked together with all other tumblers in its chamber when the cylinder plug is in the locked position.

Jiggle key

(n.) *See* "manipulation key."

Jumble cylinder

(n.) A rim or mortise cylinder of 1-1/2-inch diameter.

K

The symbol for "keys" used after a numerical designation of the quantity of the keys requested to be supplied with the cylinders. For example, 1k, 2k, 3k, and so forth. This is usually found in hardware/keying schedules.

KA

(abb.) Keyed alike.

KA1, KA2, and so forth

The symbol that indicates all cylinders so designated are or are to be operated by the same key(s). The numerical designation indicates the keyed alike group or set.

KA/2, KA/3, and so forth

The symbol used to indicate the quantity of locks or cylinders in keyed alike groups. These groups are usually formed from a larger quantity. For example, 30 cylinders KA/2.

KBA

(abb.) Key bitting array.

KD

(abb.) Keyed differently.

Key

(n.) A properly combinated device that is, or most closely resembles, the device specifically intended by the lock manufacturer to operate the corresponding lock.

Key bitting array

(n.) A matrix (graphic) display of all possible bittings for change keys and master keys as related to the top master key.

Key bitting punch

(n.) A manually operated device that stamps or punches the cuts into the key blade, rather than grinding or milling them.

Key bitting specifications

(n.) The technical data required to bit a given (family of) key blank(s) to the lock manufacturer's dimensions.

Key blank

(n.) Any material manufactured to the proper size and configuration that allows its entry into the keyway of a specific locking device. A key blank has not yet been combinated or cut.

Key bypass

(n.) and (adj.) *See* "key override."

Key change number

(n.) 1. *See* "blind code." 2. *See* "direct code." 3. *See* "key symbol."

Key changeable

(adj.) Of, or pertaining to, a lock or cylinder that can be recombinated without disassembly, by the use of a key. The use of a tool may also be required.

Key changes

(n.) 1. *See* "practical key changes." 2. *See* "theoretical key changes."

Key coding machine

(n.) A key machine designed for the production of code keys, which may or may not also serve as a duplicating machine.

Key control

(n.) 1. Any method or procedure that limits unauthorized acquisition of a key and/or controls distribution of authorized keys. 2. A systematic organization of keys and key records.

Key cult(s)

(n.) The portion of the key blade that remains after being cut and that aligns the tumbler(s).

Key cut profile

(n.) The shape of a key cut, including the cut angle and the cut root shape.

Key duplicating machine

(n.) A key machine that is designed to make copies from a pattern key.

Key gauge

(n.) A usually flat device with a cutaway portion indexed with a given set of depth or spacing specifications. This is used to help determine the combination of a key.

Key-in-knob cylinder

(n.) A cylinder used in a key-in-knob lockset.

Key interchange

(n.) An undesirable condition, usually in a master key system, whereby a key unintentionally operates a cylinder or lock.

Key machine

(n.) Any machine designed to cut keys. *See also* "key coding machine" and "key duplicating machine."

Key manipulation

(n.) Manipulation of an incorrect key to operate a lock or cylinder.

Key milling

(n.) The grooves machined into the length of the key blade to allow its entry into the keyway.

Key override

1. (n.) A provision allowing interruption or circumvention of normal operation of a combination lock or electrical device. (adj.) 2. Of, or pertaining to, such a provision, as in "key override cylinder."

Key override cylinder

(n.) A lock cylinder installed in a device to provide a key override function.

Key picking

(n.) *See* "key manipulation."

Key pin

(n.) *See* "bottom pin."

Key profile

(n.) *See* "key section."

Key pull position

(n.) Any position of the cylinder plug at which the key can be removed.

Key punch

(n.) *See* "key bitting punch."

Key records

(n.) Records that typically include some or all of the following: bitting list, key bitting array, key system schematic, end user, number of keys/cylinders issued, names of persons to whom keys were issued, and hardware/keying schedule.

Key retaining

(adj.) 1. Of, or pertaining to, a lock that must be locked before its key can be removed. 2. Of, or pertaining to, a cylinder or lock that may prevent removal of a key without the use of an additional key and/or tool.

Key section

(n.) The exact cross-sectional configuration of a key blade as viewed from the bow toward the tip.

Key stop

(n.) *See* "stop (of a key)."

Key symbol

(n.) A designation used for a key combination in the standard key coding system. For example, A, AA, AA1, and so forth.

Key system schematic

(n.) A drawing with blocks utilizing keying symbols, usually illustrating the hierarchy of all keys within a master key system. It indicates the structure and total expansion of the system.

Key trap core/cylinder

(n.) A special core or cylinder designed to capture any key to which it is combinated, once that key is inserted and turned slightly.

Keyed

(adj.) 1. Combinated. 2. Having provision for operation by key.

Keyed alike

(adj.) Of, or pertaining to, two or more locks or cylinders that have or are to have the same combination. They may or may not be part of a keying system.

Keyed common

(adj.) *See* "maison key system."

Keyed differently

(adj.) Of, or pertaining to, a group of locks or cylinders, each of which is or is to be combinated differently from the others. They may or may not be part of a keying system.

Keyed random

(adj.) Of, or pertaining to, a cylinder or group of cylinders selected from a limited inventory of different key changes. Duplicate bittings may occur.

Keying

(n.) Any specification for how a cylinder or group of cylinders are or are to be combinated to control access.

Keying chart

(n.) 1. *See* "pinning chart." 2. *See* "progression list." 3. *See* "bitting list." 4. *See* "key system schematic."

Keying conference

(n.) A meeting of the end user and the key system supplier at which the keying, and levels of keying, including future expansion, are determined and specified.

Keying diagram

(n.) *See* "key system schematic."

Keying kit

> (n.) A compartmented container that holds an assortment of tumblers, springs, and/or other parts.

Keying levels

> (n.) *See* "levels of keying."

Keying schedule

> (n.) A detailed specification of the keying system that lists how all cylinders are to be keyed, and the quantities, markings, and shipping instructions of all keys and/or cylinders to be provided.

Keying symbol

> (n.) A designation used for a lock or cylinder combination in the standard key coding system. For example, AA1, XAA1, X1X, and so forth.

Keyset

> (n.) 1. *See* "key symbol." 2. *See* "keying symbol."

Keyway

> (n.) The opening in a lock or cylinder that is shaped to accept a key bit or blade of a proper configuration.

Keyway shutter

> (n.) *See* "dust cover."

Keyway unit

> (n.) The plug of certain binary-type disc tumbler key-in-knob locks.

KR

> (abb.) 1. Keyed random. (abb.) 2. Key retaining.

KWY

> (abb.) Keyway.

layout board

> (n.) *See* "layout tray."

Layout tray

> (n.) A compartmented container used to organize cylinder parts during keying or servicing.

Lazy cam/tailpiece

> (n.) A cam or tailpiece designed to remain stationary while the cylinder plug is partially rotated and/or vice versa.

Level (of a cut)

> (n.) 1. *See* "bitting depth." 2. *See* "root depth."

Levels of keying

> (n.) The divisions of a master key system into hierarchies of access, as shown in the following tables. NOTE: the standard

key coding system has been expanded to include symbols for systems of more than four levels of keying.

Two-Level System

Level of Keying	Key Name	Abb.	Key Symbol
Level II	Master key	MK	AA
Level I	Change key	CK	1AA, 2AA, etc.

Three-Level System

Level of Keying	Key Name	Abb.	Key Symbol
Level III	Grand master key	GMK	A
Level II	Master key	MK	AA, AB, etc.
Level I	Change key	CK	AA1, AA2, etc.

Four-Level System

Level of Keying	Key Name	Abb.	Key Symbol
Level IV	Great grand master key	GGMK	GGMK
Level III	Grand master key	GMK	A, B, etc.
Level II	Master key	MK	AA, AB, etc.
Level I	Change key	CK	AA1, AA2, etc.

Five-Level System

Level of Keying	Key Name	Abb.	Key Symbol
Level V	Great-great grand master key	GGGMK	GGGMK
Level IV	Great grand master key	GGMK	A, B, etc.
Level III	Grand master key	GMK	AA, AB, etc.
Level II	Master key	MK	AAA, AAB, etc.
Level I	Change key	CK	AAA1, AAA2, etc.

Six-Level System

Level of Keying	Key Name	Abb.	Key Symbol
Level VI	Great-great grand master key	GGGMK	GGGMK
Level V	Great grand master key	GGMK	A, B, etc.
Level IV	Grand master key	GMK	AA, AB, etc.
Level III	Master key	MK	AAA, AAB, etc.
Level II	Submaster key	SMK	AAAA, AAAB, etc.
Level I	Change key	CK	AAAA1, AAAA2, etc.

Lever tumbler

A flat, spring-loaded tumbler that pivots on a post. It contains a gate that must be aligned with a fence to allow movement of the bolt.

Loading tool

(n.) A tool that aids installation of cylinder components into the cylinder shell.

Lock bumping

(v.) *See* "rap."

Lockout

(n.) Any situation in which the normal operation of a lock or cylinder is prevented.

Lockout key

(n.) A key made in two pieces. One piece is trapped in the keyway by the tumblers when inserted and it blocks entry of any regular key. The second piece is used to remove the first piece.

MACS

(abb.) Maximum adjacent-cut specification.

Maid's master key

(n.) The master key in a hotel master key system given to the maid. It operates only cylinders of the guest rooms and linen closets in the maid's designated area.

Maintenance master key

(n.) *See* "engineer's key."

Maison key system

(n.) (from the French, meaning "house" key system) A keying system in which one or more cylinders are operated by every key (or relatively large numbers of different keys) in the system. For example, main entrances of apartment buildings operated by all individual suite keys of the building.

Manipulation key

(n.) Any key other than a correct key that can be variably positioned and/or manipulated in a keyway to operate a lock or cylinder.

Master

(n.) *See* "master key."

Master blank

(n.) 1 *See* "multisection key blank." 2. *See* "all-section key blank."

Master chip

(n.) *See* "master pin."

Master disc

(n.) 1. *See* "master pin." 2. *See* "stepped tumbler." 3. A special disc tumbler with multiple gates to receive a sidebar.

Master key

(n.) 1. A key that operates all the master keyed locks or cylinders in a group, each lock or cylinder usually operated by its own change key. (v.) 2. To combinate a group of locks or cylinders, such that each is operated by its own change key, as well as by a master key for the entire group.

Master key changes

(n.) The number of different usable change keys available under a given master key.

Master key constant

(n.) *See* "constant cut."

Master key section

(n.) 1. *See* "multisection key blank." 2. *See* "all-section key blank."

Master key system

(n.) 1. Any keying arrangement that has two or more levels of keying. 2. A keying arrangement that has exactly two levels of keying.

Master keyed

(adj.) Of, or pertaining to, a cylinder or group of cylinders that are or are to be combinated, so all may be operated by their own change key(s) and by additional key(s), known as master key(s).

Master keyed only

(adj.) Of, or pertaining to, a lock or cylinder that is or is to be combinated only to a master key.

Master keying

(v.) *See* "master key."

Master lever

(n.) A lever tumbler that can align some or all other levers in its lock, so their gates are at the fence. This is typically used in locker locks.

Master pin

(n.) 1. Usually a cylindrically shaped tumbler, flat on both ends, placed between the top and bottom pin to create an additional shear line. 2. A pin tumbler with multiple gates to accept a sidebar.

Master ring

> (n.) A lube-shaped sleeve located between the plug and shell of certain cylinders to create a second shear line. Normally, the plug shear line is used for change key combinations and the shell shear line is used for master key combinations.

Master ring lock/cylinder

> (n.) A lock or cylinder equipped with a master ring.

Master wafer

> (n.) 1. *See* "master pin." 2. *See* "stepped tumbler." 3. A ward used in certain binary-type disc tumbler key-in-knob locks.

Maximum adjacent-cut differential

> (n.) *See* "maximum adjacent-cut specification."

Maximum adjacent-cut specification

> (n.) The maximum allowable difference between adjacent cut depths.

Maximum opposing-cut specifications

> (n.) The maximum allowable depths to which opposing cuts can be made without breaking through the key blade. This is typically a consideration with dimple keys.

Milling (of a key)

> (n.) 1. *See* "key section." 2. *See* "key milling."

Miscut

> (adj.) 1. Of, or pertaining to, a key that has been cut incorrectly. (n.) 2. A miscut key.

MK

> (abb.) Master key.

MK'd

> (abb.) Master keyed.

MK'd only

> (abb.) Master keyed only.

MK section

> (abb.) Master key section.

MOCS

> (abb.) Maximum opposing-cut specification.

Mogul cylinder

> (n.) A large pin tumbler cylinder whose pins, springs, key, and so forth are also proportionally increased in size. This is typically used in prison locks.

Mortise cylinder

    (n.) A threaded cylinder typically used in mortise locks of American manufacturers.

Mortise cylinder blank

    (n.) *See* "cylinder blank."

Movable constant

    (n.) *See* "rotating constant."

Multiple gating

    (n.) A means of master keying by providing a tumbler with more than one gate.

Multiplex key blank

    (n.) Any key that is part of a multiplex key system.

Multiplex key system

    (n.) 1. A series of different key sections that may be used to expand a master key system by repeating bittings on additional key sections. The keys of one key section will not enter the keyway of another key section. This type of system always includes another key section that will enter more than one, or all, of the keyways. 2. A keying system that uses such keyways and key sections.

Multisection key blank

    (n.) A key section that enters more than one keyway, but not all keyways in a multiplex key system.

Mushroom driver

    (n.) *See* "mushroom pin."

Mushroom pin

    (n.) A pin tumbler, usually a top pin, that resembles a mushroom. It is typically used to increase pick resistance.

NCK

    The symbol for "no change key," primarily used in hardware schedules.

Negative locking

    (n.) Locking achieved solely by spring pressure or gravity that prevents a key cut too deeply from operating a lock or cylinder.

NKR

    (abb.) Nonkey retaining.

NMK

    A keying symbol that means "not master keyed" and is suffixed in parentheses to the regular key symbol. It indicates the cylinder is not to be operated by the master key(s) specified in the regular key symbol. For example, AB6 (NMK).

Nonkey retaining

(adj.) Of, or pertaining to, a lock whose key can be removed in both the locked and unlocked positions.

Nonkeyed

(adj.) Having no provision for key operation. NOTE: This term also includes privacy function locksets operated by an emergency key.

Nonoriginal key blank

(n.) Any key other than an original key blank.

O billed

(adj.) *See* "zero bitted."

Odometer method

(n.) A means of progressing key bittings using a progression sequence of right to left.

One bitted

(adj.) Of, or pertaining to, a cylinder that is or is to be combinated to keys cut to the manufacturer's reference number-one bitting.

One-column progression

(n.) A process wherein key bittings are obtained by using the cut possibilities in one column of the key bitting array.

One-pin master key

(n.) A master key for all combinations obtained by progressing only one bitting position.

Open code

(n.) *See* "direct code."

Operating key

(n.) 1. Any key that will properly operate a lock or cylinder to lock or unlock the lock mechanism and is not a control key or a reset key. 2. *See* "change key."

Original key

(n.) 1. *See* "factory original key." 2. *See* "code original key."

Original key blank

(n.) A key blank supplied by the lock manufacturer to fit that manufacturer's specific product.

Page master key

(n.) The three-pin master key for all combinations listed on a page in the standard progression format.

Paracentric

(adj.) 1. Of, or pertaining to, a keyway with one or more wards on each side projecting beyond the vertical center line of the keyway to hinder picking. 2. Of, or pertaining to, a key blank made to enter such a keyway.

Pattern key

(n.) 1. An original key kept on file to use in a key duplicating machine when additional keys are required. 2. Any key that is used in a key duplicating machine to create a duplicate key.

Peanut cylinder

(n.) A mortise cylinder of 3/4" diameter.

Phantom key

(n.) *See* "incidental master key."

Pick

(n.) 1. A tool or instrument, other than the specifically designed key, made for the purpose of manipulating tumblers in a lock or cylinder into the locked or unlocked position through the keyway, without obvious damage. (v.) 2. To manipulate tumblers in a keyed lock mechanism through the keyway, without obvious damage, by means other than the specifically designed key.

Pick key

(n.) 1. A type of manipulation key, cut or modified to operate a lock or cylinder pin. 2. *See* "pin tumbler." (v.) 3. To install pin tumblers into a cylinder and/or cylinder plug.

Pin cell

(n.) *See* "pin chamber."

Pin chamber

(n.) The corresponding hole drilled into the cylinder shell and/or plug to accept the pin(s) and spring.

Pin kit

(n.) A type of keying kit for a pin-tumbler mechanism.

Pin segment

(n.) *See* "pin tumbler."

Pin set

(v.) *See* "pin."

Pin slack

(n.) 1. All the tumblers in a given pin chamber. 2. *See* "pin stack height."

Pin slack height

> (n.) The measurement of a pin stack, often expressed in units of the lock manufacturer's increment or as an actual dimension.

Pin tray

> (n.) *See* "layout tray."

Pin tumbler

> (n.) Usually a cylindrically shaped tumbler. Three types are normally used: bottom pin, master pin, and top pin.

Pin tweezers

> (n.) A tool used in handling tumblers and springs.

Pinning

> (v.) *See* "pin."

Pinning block

> (n.) 1. A holding fixture that assists in the loading of tumblers into a cylinder or cylinder plug pinning chart. 2. A numerical diagram that indicates the sizes and order of installation of the various pins into a cylinder. The sizes are usually indicated by a manufacturer's reference number that equals the quantity of increments a tumbler represents.

Plug

> (n.) The part of a cylinder that contains the keyway, with tumbler chambers usually corresponding to those in the cylinder shell.

Plug follower

> (n.) A tool used to allow removal of the cylinder plug while retaining the top pins, springs, and/or other components within the shell.

Plug holder

> (n.) A holding fixture that assists in the loading of tumblers into a cylinder plug.

Plug iron

> (n.) 1. *See* "plug follower." 2. *See* "set-up plug."

Plug retainer

> (n.) The cylinder component that secures the plug in the shell.

Plug set-up chart

> (n.) *See* "pinning chart."

Plug vise

> (n.) *See* "plug holder."

Positional master keying

(n.) A method of master keying typical of certain binary type's disc tumbler key-in-knob locks and of magnetic and dimple key cylinders. Of all possible tumbler positions within a cylinder, only a limited number contain active tumblers. The locations of these active tumblers are rotated among all possible positions to generate key changes. Higher level keys must have more cuts or magnets than lower level keys.

Positive locking

(n.) The condition brought about when a key cut that is too high forces its tumbler into the locking position. This type of locking does not rely on gravity or spring pressure.

Practical key changes

(n.) The total number of usable different combinations available for a specific cylinder or lock mechanism.

Prep key

(n.) A type of guard key for a safe deposit box lock with only one keyway. It must be turned once and withdrawn before the renter's key will unlock the unit.

Privacy key

(n.) 1. A key that operates an SKD cylinder. 2. *See* "emergency key."

Profile

(n.) 1. *See* "key section." 2. *See* "keyway."

Profile cylinder

(n.) A cylinder with a usually uniform cross section, which slides into place and is held by a mounting screw. It is typically used in mortise locks of non-U.S. manufacturers.

Program key

(n.) *See* "reset key."

Progress

(v.) To select possible key bittings from the key bitting array, usually in numerical order.

Progression

(n.) A logical sequence of selecting possible key bittings, usually in numerical order from the key bitting array.

Progression column

(n.) A listing of the key bitting possibilities available in one bitting position, as displayed in a column of the key bitting array.

Progression formula

(n.) *See* "key bitting array."

Progression list

> (n.) A bitting list of change keys and master keys arranged in sequence of progression.

Progression sequence

> (n.) *See* "sequence of progression."

Progressive

> (n.) Any bitting position that is progressed, rather than held constant.

Proprietary

> (adj.) Of, or pertaining to, a keyway and key section assigned exclusively to one end user by the lock manufacturer. It may also be protected by law from duplication.

Quadrant master key

> (n.) *See* "four-pin master key."

Radiused blade bottom

> (n.) The bottom of a key blade that has been radiused to conform to the curvature of the cylinder plug it is designed to enter.

Random master keying

> (v.) To rekey by installing a different core.

Rap

> (v.) 1. [Also called "lock bumping."] To unlock a plug from its shell by striking sharp blows to the spring of the cylinder while applying tension to the plug. 2. To unlock a padlock from its shell by striking sharp blows to the sides to disengage the locking dogs.

Read key

> (n.) A key that allows access to the sales and/or customer data on certain types of cash control equipment (for example, cash registers).

Recode

> (v.) *See* "recombinate."

Recombinate

> (v.) To change the combination of a lock, cylinder, or key.

Recore

> (v.) To rekey by installing a different core.

Register groove

> (n.) The reference point on the key blade from which some manufacturers locate the hitting depths.

Register number

(n.) 1. A reference number, typically assigned by the lock manufacturer to an entire master key system. 2. A blind code assigned by some lock manufacturers to higher level keys in a master key system.

Rekey

(v.) To change the existing combination of a cylinder or lock.

Removable core

(n.) A key removable core that can only be installed in one type of cylinder housing. For example, a rim cylinder or mortise cylinder, or a key-in-knob lock.

Removable cylinder

(n.) A cylinder that can be removed from a locking device by a key and/or tool.

Removal key

(n.) 1. The part of a two-piece key that is used to remove its counterpart from a keyway. 2. *See* "control key" #1 and #3. 3. *See* "construction breakout key."

Renter's key

(n.) A key that must be used together with a guard key, prep key, or electronic release to unlock a safe deposit lock. This key is usually different for every unit within an installation.

Repin

(v.) To replace pin tumblers, with or without changing the existing combination.

Reserved

(adj.) *See* "restricted."

Reset

(v.) *See* "recombinate."

Reset key

(n.) 1. A key used to set some types of cylinders to a new combination. Many of these cylinders require the additional use of tools and/or the new operating key to establish the new combination. 2. A key that allows the tabulations on various types of cash control equipment (for example, cash registers) to be cleared from the records of the equipment.

Restricted

(adj.) Of, or pertaining to, a keyway and corresponding key blank whose sale and/or distribution is limited by the lock manufacturer to reduce unauthorized key proliferation.

Retainer

(n.) *See* "plug retainer."

Reversible key

(n.) A symmetrical key that may be inserted either way up to operate a lock.

Rim cylinder

(n.) A cylinder typically used with surface applied locks, and attached with a back plate and machine screws. It has a tailpiece to actuate the lock mechanism.

Rocker key

(n.) *See* "manipulation key."

Root depth

(n.) The dimension from the bottom of a cut on a key to the bottom of the blade.

Root of cut

(n.) *See* "cut root."

Rose

(n.) A usually circular escutcheon.

Rotary tumbler

(n.) A circular tumbler with one or more gates. Rotation of the proper key aligns the tumbler gates at a sidebar, fence, or shackle slot.

Rotating constant

(n.) One or more cut(s) in a key of any level that remain constant throughout all levels and are identical to the top master key cuts in their corresponding positions. The positions where the top master key cuts are held constant may be moved, always in a logical sequence.

Rotating constant method

(n.) A method used to progress key bittings in a master key system, wherein at least one cut in each key is identical to the corresponding cut in the top master key. The identical cut(s) is moved to different locations in a logical sequence until each possible planned position has been used.

Row master key

(n.) The one pin master key for all combinations listed on the same line across a page in the standard progression format.

S/A

(abb.) Subassembled.

Sample key

(n.) *See* "pattern key."

Scalp

(n.) A thin piece of metal that is usually crimped or spun onto the front of a cylinder. It determines the cylinder's finish and may also serve as the plug retainer.

Schematic

(n.) *See* "key system schematic."

Second-generation duplicate

(n.) A key reproduced from a first-generation duplicate.

Sectional key blank

(n.) *See* "multiplex key blank."

Sectional keyway system

(n.) *See* "multiplex key system."

Security collar

(n.) 1. A protective cylinder collar. 2. *See* "cylinder guard."

Segmented follower

(n.) A plug follower that is sliced into sections, which are introduced into the cylinder shell one at a time. It is typically used with profile cylinders.

Selective key system

(n.) A key system in which every key has the capability of being a master key. It is normally used for applications requiring a limited number of keys and extensive cross keying.

Selective master key

(n.) An unassociated master key that can be made to operate any specific lock(s) in the entire system in addition to the regular master key(s) and/or change key(s) for the cylinder, without creating key interchange.

Sequence of progression

(n.) The order in which bitting positions are progressed to obtain change key combinations.

Series water

(n.) A type of disc tumbler used in certain binary type disc tumbler key-in-knob locks. Its presence requires that no cut be made in that position on the operating key(s).

Set

(v.) *See* "combinate."

Set-up key

(n.) A key used to calibrate some types of key machines.

Set-up plug

(n.) A type of loading tool shaped like a plug follower. It contains pin chambers and is used with a shove knife to load springs and top pins into a cylinder shell.

Seven-column progression

(n.) A process wherein key bittings are obtained by using the cut possibilities in seven columns of the key bitting array.

Seven-pin master key

(n.) A master key for all combinations obtained by progressing seven bitting positions.

Shaved blade

(n.) *See* "radiused blade bottom."

Shear line

(n.) A location in a cylinder at which specific tumbler surfaces must be aligned, removing obstruction(s) that prevented the plug from moving.

Shedding key

(n.) *See* "declining step key."

Shell

(n.) The part of the cylinder that surrounds the plug and usually contains tumbler chambers corresponding to those in the plug.

Shim

(n.) 1. A thin piece of material used to unlock the cylinder plug from the shell by separating the pin tumblers at the shear line, one at a time. (v.) 2. To unlock a cylinder plug from its shell by using a shim.

Shoulder

(n.) 1. Any key stop other than a tip stop. 2. *See* "bow stop."

Shouldered pin

(n.) A bottom pin whose diameter is larger at the flat end to limit its penetration into a counter-bored chamber.

Shove knife

(n.) A tool used with a set-up plug that pushes the springs and pin tumblers into the cylinder shell.

Shut-out key

(n.) Usually used in hotel keying systems, a key that will make the lock inoperative to all other keys in the system, except the emergency master key, display key, and some types of shut-out keys.

Shut-out mode

(n.) The state of a hotel function lockset that prevents operation by all keys, except the emergency master key, display key, and some types of shut-out keys.

Shutter

(n.) *See* "dust cover."

Sidebar

(n.) A primary or secondary locking device in a cylinder. When locked, it extends along the plug beyond its circumference. It must enter gates in the tumblers to clear the shell and allow the plug to rotate.

Simplex key section

(n.) A single, independent key section that cannot be used in a multiplex key system.

Single key section

(n.) An individual key section that can be used in a multiplex key system.

Single-step progression

(n.) A progression using a one-increment difference between billings of a given position.

Six-column progression

(n.) A process wherein key bittings are obtained by using the cut possibilities in six columns of the key bitting array.

Six-pin master key

(n.) A master key for all combinations, obtained by progressing six bitting positions.

SKD

Symbol for "single keyed," normally followed by a numerical designation in the standard key coding system. For example, SKD1, SKD2, and so forth. It indicates a cylinder or lock is not master keyed, but is part of the keying system.

Skew

(n.) *See* "degree of rotation."

Slide

(n.) *See* "spring cover."

SMK

(abb.) Submaster key.

Spacing

(n.) The dimensions from the stop to the center of the first cut and/or to the centers of successive cuts.

Special application cylinder

> (n.) Any cylinder other than a mortise, rim, key-in-knob, or profile cylinder.

Split pin

> (n.) *See* "master pin."

Split pin master keying

> (n.) A method of master keying in a pin tumbler cylinder by installing master pins into one or more pin chambers.

Spool pin

> (n.) Usually a top pin that resembles a spool, typically used to increase pick resistance.

Spring cover

> (n.) A device for sealing one or more pin chambers.

Stack height

> (n.) *See* "pin stack height."

Standard key coding system

> (n.) An industry standard and uniform method of designating all keys and/or cylinders in a master key system. The designation automatically indicates the exact function and keying level of each key and/or cylinder in the system, usually without further explanation.

Standard progression formal

> (n.) A systematic method of listing and relating all change key combinations to all master key combinations in a master key system. The listing is divided into segments known as blocks, horizontal groups, vertical groups, rows, and pages for levels of control.

Step

> (n.) *See* "increment."

Step pin

> (n.) A spool or mushroom pin that has had a portion of its end machined to a smaller diameter than the opposite end. It is typically used as a top pin to improve pick resistance by some manufacturers of high-security cylinders.

Step tolerance

> (n.) *See* "maximum adjacent cut specification."

Stepped tumbler

> (n.) A special (usually disc) tumbler used in master keying. It has multiple bearing surfaces for blades of different key sections.

Stop (of a key)

    (n.) *See* "maximum adjacent-cut specification."

Subassembled

    (adj.) *See* "uncombinated."

Submaster key

    (n.) The master key level immediately below the master key in a system of six or more levels of keying.

Tailpiece

    (n.) An actuator attached to the rear of the cylinder, parallel to the plug, typically used on rim, key-in-knob, or special purpose cylinders.

Template keys

    (n.) *See* "depth key set."

Theoretical key changes

    (n.) The total possible number of different combinations available (or a specific cylinder or lock mechanism).

Thimble

    (n.) *See* "plug holder."

Threaded cylinder

    (n.) *See* "mortise cylinder."

Three-column progression

    (n.) A process wherein key billings are obtained by using the cut possibilities in three columns of the key bitting array.

Three-pin master key

    (n.) A master key for all combinations obtained by progressing three bitting positions.

Thumb turn cylinder

    (n.) A cylinder with a turn knob, rather than a keyway and a tumbler mechanism.

Tip

    (n.) The portion of the key that enters the keyway first.

Tip stop

    (n.) A type of stop located at or near the tip of the key.

Tolerance

    (n.) The deviation allowed from a given dimension.

Top master key

    (n.) The highest level master key in a master key system.

Top of blade

    (n.) The bitted edge of a single bitted key.

Top pin

(n.) Usually a cylindrically shaped tumbler, flat on both ends and installed directly under the spring in the pin stack.

Total position progression

(n.) A process used to obtain key bittings in a master key system, wherein billings of change keys differ from those of the top master key in all bitting positions.

Total stack height

(n.) *See* "pin stack."

Trim ring

(n.) 1. *See* "cylinder collar." 2. *See* "rose."

Try-out key

(n.) A manipulation key that is usually part of a set, used for a specific series, keyway, and/or brand of lock.

Tubular key

(n.) A key with a tubular blade. The key cuts are made into the end of the blade, around its circumference.

Tumbler

(n.) A movable obstruction of varying size and configuration in a lock or cylinder. It makes direct contact with the key or another tumbler and prevents an incorrect key or torquing device from activating the lock or other mechanism.

Tumbler spring

(n.) Any spring that acts directly on a tumbler.

Two-column progression

(n.) A process wherein key bittings are obtained by using the cut possibilities in two columns of the key bitting array.

Two-pin master key

(n.) A master key for all combinations obtained by progressing two bitting positions.

Two-step progression

(n.) A progression using a two-increment difference between bittings of a given position.

UL

(abb.) Underwriters Laboratories.

Unassociated change key

(n.) A change key that is not related directly to a particular master key through the use of certain constant cuts.

Unassociated master key
> (n.) A master key that does not have change keys related to its combination through the use of constant cuts.

Uncoded
> (adj.) *See* "uncombinated."

Uncombinated
> (adj.) 1. Of, or pertaining to, a cylinder that is or is to be supplied without keys, tumblers, and springs. 2. Of, or pertaining to, a lock, cylinder or key in which the combination has not been set.

Uncontrolled cross keying
> (n.) A condition in which two or more different keys under *different* higher level keys operate one cylinder by design. For example, XAA1 operated by AB, AB1, and so forth. NOTE: This condition severely limits the security of the cylinder and the maximum expansion of the system. It often leads to key interchange.

Unidirectional cylinder
> (n.) A cylinder whose key can turn in only one direction from the key pull position, often not making a complete rotation.

Universal keyway
> (n.) *See* "composite keyway."

Vertical group master key
> (n.) The two-pin master key for all combinations listed in all blocks in a line down a page in the standard progression format.

VGM
> (abb.) Vertical group master key.

Visual key control
> (n.) A specification that all keys and the visible portion of the front of all lock cylinders be stamped with standard keying symbols.

VKC
> (abb.) Visual key control.

Wafer
> (n.) 1. *See* "disc tumbler." 2. *See* "master pin."

Ward
> (n.) A usually stationary obstruction in a lock or cylinder that prevents the entry and/or operation of an incorrect key.

Ward cut
> (n.) A modification of a key that allows it to bypass a ward.

Wiggle key

> (n.) *See* "manipulation key."

Zero billed

> (adj.) Of, or pertaining to, a cylinder that is or is to be combinated to keys cut to the manufacturer's reference number 0 bitting.

Index

BHMA (Builders Hardware
    Manufacturers Association), 11
BHMA finish codes, 257–268
bible, 359
bicentric cylinder, 359
bidirectional cylinder, 359
binary cut keys, 359
binary type cylinder/lock, 359
biometric systems, 201, 202
bit, 359
bit-key locks, 7–8
bitting, 359
blade, 359
blank face hinges, 94
blanks. *See* key blanks
Blaydes Industries Inc., 297–298
blind code, 360
block master keys, 360
Blue Dog, 298
bolts. *See also* deadbolts
    emergency exit doors, 111, 112
    lock checklist, 11
    lockbolt, 239
bonding, 244–246
books, writing, 263–264
bored-in deadbolts, 3–4
bottom pin, 360
bow, 360
bow stop, 360
Boyle & Chase Inc., 298
broach, 360
brochures, 234, 266
Builders Hardware Manufacturers
    Association. *See* BHMA
building master keys, 360
bump keys, 19–30
burglaries. *See also* alarm systems
    bump keys and, 19–20, 23, 29
    discouraging, 256–259
    piggybacking and, 259
burglary safe ratings, 230–231
burglary safes, 225–226, 229–231. *See
    also* safes
business cards, 234, 266
butts, door, 93–96
bypass keys, 360

California Security Alarms, Inc., 336
calipers, 27, 34
cam, 361
cam lock, 361

cameras
    apartment access and, 202–203
    CCTV systems, 197–199
    home security, 197–199, 257
    security systems, 197–199
cap, 361
capacitors, 109, 326
Capitol Lock Co., Inc., 299
capping block, 361
car locks. *See* automobile keys/locks
card systems, 201
C.C. Craig Co. Ltd., 298
CCTV (closed-circuit television) system,
    197–199
cctvproducts.com, 299
Central District Alarm, Inc., 336
Central Lock & Hardware Supply Co., 299
central monitoring stations, 175, 335–341
Central Station Monitoring, 336
certification tests, xv, 244–245
Certified Alarm Distributors, 299
certified protection professional (CPP)
    test, xv
chamber, 361
Chamber of Commerce, 252
change keys, 224, 358, 361, 398
changeable bit keys, 361
Christy Industries Inc, 299
circuits, 102–110, 162
City Lock, 299–300
Clark Security Products, 300–301
Class 6 government containers, 238
clear swing hinge, 95–96
ClearStar.com web site, xv, 238, 240
closed-circuit television (CCTV) system,
    197–199
clutch, 362
CMT Inc., 302
code, 362
code keys, 362
code original keys, 362
coding systems, 215
coils (inductors), 329
Colonial Lock Supply Co. Inc., 302
combinate, 362
combination, 362
combination circuit, 102
combination dial safes, 229, 238–239
combination water, 362
Commonwealth Lock Co., 302
compensate drivers, 363